SCIENTIFIC AMERICAN EXPLORES BIG IDEAS

Genetic Engineering

The Editors of *Scientific American*

SCIENTIFIC AMERICAN EDUCATIONAL PUBLISHING

New York

Published in 2024 by Scientific American Educational Publishing
in association with **The Rosen Publishing Group**
2544 Clinton Street, Buffalo NY 14224

Contains material from Scientific American®, a division of Springer Nature America, Inc., reprinted by permission, as well as original material from The Rosen Publishing Group®.

First Edition

Scientific American
Lisa Pallatroni: Project Editor

Rosen Publishing
Daniel R. Faust: Compiling Editor
Michael Moy: Senior Graphic Designer

Cataloging-in-Publication Data
Names: Scientific American Editors (New York, New York), editor.
Title: Genetic engineering / The Editors of Scientific American.
Description: First edition. | New York : Scientific American Educational Publishing, 2024. | Series: Scientific American explores big ideas | Includes bibliographical references and index. | Audience: Grades 10-12
Identifiers: LCCN 2023036671 | ISBN 9781725350434 (library binding) | ISBN 9781725350427 (paperback) | ISBN 9781725350441 (ebook)
Subjects: LCSH: Genetic engineering–Juvenile literature.
Classification: LCC QH442 .G44319 2024 | DDC 576.5–dc23/eng/20230825
LC record available at https://lccn.loc.gov/2023036671

Manufactured in the United States of America
Websites listed were live at the time of publication.

Cover: KateStudio/Shutterstock.com

CPSIA Compliance Information: Batch # CWSA24.
For Further Information contact Rosen Publishing at 1-800-237-9932.

CONTENTS

INTRODUCTION

Genetic engineering has long been a common trope in popular culture, beginning with 19th-century works of science fiction such as Mary Shelley's *Frankenstein* and H.G. Wells's *The Island of Dr. Moreau*. Modern audiences marvel at the cloned dinosaurs of the *Jurassic Park* franchise and the genetically engineered monsters in video games such as the *Resident Evil* series. Advances in technology have made processes such as genetic engineering, gene manipulation, and cloning science fact instead of science fiction. Today, scientists can manipulate DNA for any number of reasons, from fighting disease to improving food production.

The articles in this collection look at the modern science of genetic engineering, including some of the most recent advances in the field. In Section 1, "Genetic Engineering in the 21st Century," readers are introduced to some of the most recent research and discoveries in genetic engineering. Section 2, "Fighting Disease," explores the ways that genetic engineering is being used in the medical field. From gene therapy to designer DNA, these advancements are looking deep inside the building blocks of life to find new ways to treat diseases and other health issues. Are foods modified by genetic engineering safe? What are the benefits of using genetics to make crops heartier? These are questions at the heart of Section 3, "What's the Big Deal About GMOs?" Section 4, "Saving Endangered Species," explores the ways that genetic engineering can be used to save endangered species and whether it can even bring species of plants and animals back from extinction. Any new technology will understandably be met with questions and concerns. Section 5, "Questions and Controversies," addresses some of these points of contention, such as the ethical implications of changing the DNA of a living thing.

Section 1: Genetic Engineering in the 21st Century

Gene Therapy's Second Act

By Ricki Lewis

Gene therapy may finally be living up to its early promise. In the past six years the experimental procedure for placing healthy genes wherever they are needed in the body has restored sight in about 40 people with a hereditary form of blindness. Doctors have seen unprecedented results among another 120-plus patients with various cancers of the blood—several of whom remain free of malignancy three years after treatment. Researchers have also used gene therapy to enable a few men with hemophilia, a sometimes fatal bleeding disorder, to go longer without dangerous incidents or the need for high doses of clotting drugs.

The positive results are even more impressive considering that the field of gene therapy essentially ground to a halt 15 years ago, following the untimely death of Jesse Gelsinger, a teenager with a rare digestive disorder. Gelsinger's immune system reacted to the gene treatment he received by launching a counterattack of unexpected ferocity that killed him. Gene therapy's preliminary successes in the 1990s, it turns out, had fueled unreasonably high expectations among doctors and researchers—and perhaps a bit of hubris.

This and other setbacks forced scientists to rethink some of their approaches, as well as to be more realistic about gene therapy's feasibility for treating various conditions in people. Investigators curbed their hopes and returned to basic research. They examined potentially fatal side effects such as those experienced by Gelsinger and learned how to avoid them. And they paid more attention to explaining the risks and benefits to volunteers and their families.

The turning point, in the view of many observers, came six years ago, when doctors treated then eight-year-old Corey Haas for a degenerative eye disorder that caused his sight to deteriorate. The gene therapy they used allowed the defective retina of Haas's left eye to make a protein that his body could not otherwise produce. Within four days he took a trip to the zoo and found, to his delight and astonishment, that he could see the sun and a hot-air balloon. Three years later he underwent the same treatment in his right eye. Now Haas sees well enough to go turkey hunting with his grandfather.

Although gene therapy is still not available in hospitals and clinics, that is likely to change in the next decade. Europe approved its first gene treatment, for a rare but extremely painful disorder called familial lipoprotein lipase deficiency, in 2012. At the end of 2013 the National Institutes of Health removed some of the regulatory speed bumps that the agency now considers unnecessary. The first U.S. approval of a commercial gene treatment, some industry watchers predict, may come in 2016. Gene therapy, after its lost decade, is at last beginning to fulfill its destiny as a revolutionary medical treatment.

Heartbreak

The early failures of gene therapy highlight how difficult it is to establish a safe and efficient means of delivering genes to the target tissue. Too often the safest delivery systems were not very effective, and some of the most effective systems turned out not to be very safe, setting off either an overwhelming immune reaction, as in Gelsinger's case, or the development of leukemia, as in other instances.

To understand what triggered these side effects and to figure out how to lessen the risks of their occurrence, scientists focused on the most common delivery system for gene therapy: engineering a virus to act as a kind of microscopic injection gun.

For starters, researchers remove some of the virus's own genes to create room for the healthy genes that they want to deliver to a patient. (This step also has the added benefit of preventing the virus from making copies of itself once inside the body, which increases the chances of an immune reaction.) Then the customized viruses are injected into that person, where they insert the new genes into various places in cells, depending on the type of virus being used.

By the time Gelsinger volunteered for a clinical trial, the delivery system of choice consisted of adenoviruses, which in their natural state can cause mild upper respiratory infections in people. Scientists at the University of Pennsylvania determined that the best chance for success was to inject the viruses into the liver, where the cells that normally make the digestive enzyme Gelsinger was missing are located. They packaged a working copy of the gene for that enzyme into stripped-down adenoviruses. Then they injected one trillion of these viruses—each with their custom payload—directly into Gelsinger's liver.

Once in Gelsinger's body, however, some of the viruses took a tragic detour. They entered the liver cells as planned, but they also infected huge numbers of macrophages, the large wandering cells that serve as sentries for the immune system, and the dendritic cells that announce an invasion. The immune system responded by destroying each infected cell, a violent process that ultimately ravaged Gelsinger's body from the inside out.

The ferocity of the immune response took investigators by surprise. None of the 17 volunteers who had previously undergone treatment for the same disorder had exhibited such severe side effects. Researchers knew that adenoviruses could provoke an immune response, but apart from a study of a slightly different reengineered virus in which a monkey died, they did not realize how explosive the reactions could be. "Humans are much more heterogeneous than colonies of animals," says James Wilson of the University of Pennsylvania, who developed the viral delivery system used in the clinical trial in which Gelsinger had participated. "What we saw in that trial was one individual out of 18 who had a very exaggerated host response." In hindsight, it seemed that it would have been wiser to inject fewer—billions rather than one trillion—gene-bearing viruses into his body. The researchers were also criticized for not informing Gelsinger and his family about the monkey's death so that they could make up their own minds about whether it was an unrelated event.

Gelsinger's death was not the only gene therapy tragedy. Soon after, treatment for another disorder—called severe combined immunodeficiency X1, or SCID-X1—triggered five cases of leukemia, including one death, in 20 children. Once again the gene delivery system turned out to be at fault. In this instance, however, the microscopic injection gun in question consisted of a retrovirus, a kind of virus that inserts its genetic payload directly into the DNA of a cell. The exact placement of the therapeutic genes is a bit haphazard, however, and the retrovirus sometimes inserted its payload into an oncogene—a gene that can cause cancer under certain circumstances.

Rethinking the Technology

Given the propensity of adenoviruses to provoke lethal immune reactions and of retroviruses to trigger cancer, investigators began paying more attention to other viruses to see if they offered better results. They soon focused on two more widely suitable entrants.

The first new delivery system, adeno-associated virus (AAV), does not make people sick (although most of us have been infected by it at one time or another). Because it is so common, it is unlikely to cause extreme immune reactions. This virus has another feature that should also help minimize side effects: it is available in several varieties, or serotypes, that favor specific types of cells or tissues. For example, AAV2 works well in the eye, whereas AAV8 prefers the liver, and AAV9 slips into heart and brain tissue. Researchers can choose the best AAV for a specific body part, decreasing the number of individual viruses that need to be injected and thus minimizing the chances of an overwhelming immune response or other unwanted reaction. Plus, AAV deposits its genetic payload outside the chromosomes, so it cannot accidentally cause cancer by interfering with oncogenes.

Adeno-associated virus was first used in a clinical trial in 1996, on cystic fibrosis. Since then, 11 serotypes have been identified, and their parts have been mixed and matched to engineer hundreds of seemingly safe and selective delivery tools. Current studies are evaluating AAV-borne gene therapy for several brain diseases, including Parkinson's and Alzheimer's, and for hemophilia, muscular dystrophy, heart failure and blindness.

The second, rather more surprising new gene vector is a stripped-down version of HIV–the virus that causes AIDS. Once you look beyond HIV's reputation as a killer, its advantages for gene therapy emerge. As a member of the *Lentivirus* genus of retroviruses, it evades the immune system and–crucial for a retrovirus–does not typically disturb oncogenes.

After the genes that make HIV lethal are removed, the viral packaging that remains "has a large capacity," says Stuart Naylor, formerly chief scientific officer at Oxford Biomedica in England, which is pursuing "gene-based medicines" for eye diseases. Unlike the smaller AAV, "it's great for installing multiple genes or big, chunky genes," he says. "There's no toxicity and no adverse immune reaction." Stripped-down lentiviruses are now being used in a number of clinical trials, including treatments for adrenoleukodystrophy–the disease featured in the 1992 movie *Lorenzo's Oil*. To date, a few of the boys who have received this treatment have become healthy enough to return to school.

Although clinical trials using AAV and HIV are on the rise, researchers have also redirected or modified the older viral delivery systems so that they can be used in limited circumstances. For example, non-HIV retroviruses are now genetically edited so that they inactivate themselves before they can trigger leukemia.

Even adenovirus, which caused Gelsinger's death, is still in clinical trails as a gene therapy vector. Investigators restrict its use to parts of the body where it is unlikely to cause an immune response. One promising application is to treat "dry mouth" in patients undergoing radiation for head and neck cancer, which damages the salivary glands, located just under the surface of the inside of the cheek.

The NIH is running a small clinical trial that involves inserting a gene that creates channels for water into the glands. Because the glands are small and contained, and the experimental design calls for 1,000-fold fewer viruses than were used on Gelsinger, the chances of an immune overreaction are reduced. In addition, viruses that do not hit their target cells should wind up in a patient's drool, either swallowed or spit out, with little chance of irking the immune system. Since 2006, six of 11 treated patients have been shown to produce significantly more saliva. Bruce Baum, a dentist and biochemist who led the research before he retired, calls the results "cautiously encouraging."

New Targets

Emboldened by these successes, medical researchers have moved beyond treating hereditary diseases to trying to reverse genetic damage that naturally occurs over the course of a lifetime.

Scientists at the University of Pennsylvania, for example, are using gene therapy to tackle a common childhood cancer known as acute lymphoblastic leukemia (ALL).

Although most children with ALL respond to standard chemotherapy, about 20 percent do not. Researchers are turning to gene therapy to turbocharge these children's immune cells to seek out and destroy the recalcitrant cancer cells.

The experimental approach is particularly complex and is based on so-called chimeric antigen receptor (CAR) technology. Like the chimera of Greek mythology that is made up of different animals, a chimeric antigen receptor consists of two molecules from the immune system that are not normally found together. Some immune cells, known as T cells, are then outfitted with these chimeric antigen receptors, which allow the cells to target proteins that are found in greater numbers on a leukemia cell. The fully armed and deployed T cell then destroys the cancer cell. The first test subjects were adults with chronic leukemia, who responded favorably. The next attempt, with a child, exceeded the researchers' wildest dreams.

Emily Whitehead was five in May 2010, when she was diagnosed with leukemia. Two rounds of chemotherapy did not work. In the spring of 2012 "she was given a [third] chemotherapy dose that would have killed an adult, and she still had lesions in her kidneys, liver and spleen," says Bruce Levine, one of Whitehead's doctors. The girl was days from death.

Doctors took a sample of Whitehead's blood and isolated some of her T cells. They then injected the sample with lentiviruses that had been outfitted with the appropriate genes. After a rocky start, which fortunately responded to treatment, Whitehead quickly improved. Three weeks after treatment, a quarter of the T cells in her bone marrow bore the genetic correction. Her T cells began homing in on the cancer cells, which soon vanished. "In April she had been bald," Levine recalls. "By August she went to her first day of second grade."

Although Whitehead's modified cells might not last forever–in which case doctors can repeat the treatment–this beautiful girl with shaggy brown hair has been free of cancer for about two years. And she is not alone. By late 2013 several groups of researchers reported that they had used the CAR technique on more than 120 patients, for Whitehead's form of leukemia and three other blood cancers. Five adults and 19 of 22 children have achieved remission, meaning that they are currently cancer-free.

Into the Clinic

With safer viral delivery systems in hand, gene therapy specialists are now tackling the greatest challenge that any new drug faces: earning the approval of the U.S. Food and Drug Administration. This daunting step requires so-called phase III clinical trials, which are designed to assess efficacy in a larger group of volunteer patients and typically take one to five years to complete (the time varies widely). As of the end of 2013, about 5 percent of approximately 2,000 clinical trials for gene therapy had reached phase III. One of the furthest along is aimed at Leber congenital amaurosis–the condition that was robbing Haas of his sight. So far several dozen patients have had corrective genes inserted into both eyes and are now able to see the world.

China was the first country to approve a gene treatment, in 2004, for head and neck cancer. In 2012 Europe approved a gene therapy-based drug called Glybera to treat familial lipoprotein lipase deficiency. Working copies of the mutant gene wrapped in AAV are injected into the leg muscles. Netherlands-based company UniQure is in early talks with the FDA about approval in the U.S. One potential stumbling block: the price tag for a single curative dose is $1.6 million, but that cost may come down as researchers develop more efficient procedures.

As with many medical technologies, the decades-long path to successful gene therapy has been circuitous and is far from complete. But as gene therapy accumulates more success stories such as Corey Haas and Emily Whitehead, it is moving closer to becoming a mainstream medical treatment for some disorders and a promising new option for others.

Referenced

Gene Therapy of Inherited Retinopathies: A Long and Successful Road from Viral Vectors to Patients. Pasqualina Colella and Alberto Auricchio in *Human Gene Therapy*, Vol. 8, No. 23, pages 796–807; August 2012. www.ncbi.nlm.nih.gov/pubmed/22734691

National Institutes of Health's gene therapy website: http://ghr.nlm.nih.gov/handbook/therapy

About the Author

Ricki Lewis is a science writer with a Ph.D. in genetics. She is author of several textbooks, many magazine articles and the book The Forever Fix: Gene Therapy and the Boy Who Saved It *(St. Martin's Press, 2012).*

What Gene Therapy Needs Now: A Good Off Switch

By Jim Kozubek

Humans don't molt," R.J. Kirk tells me. Kirk is a billionaire geek who runs his offices out of West Palm Beach, Fla., a balmy land of pelicans and tangled mangroves. He built his fortune on conventional medications that can be taken as a pill, and I had phoned to talk about his newest endeavors in biotech. I wasn't expecting to hear about bugs. But the molting process, in which a growing insect builds a new exoskeleton to replace an old one that no longer fits, turns out to have some very important properties that can be adapted to make gene therapy, still a largely experimental procedure, safer.

Doctors would like to deliver copies of working genes to people to treat a variety of hereditary ills. Genes provide cells with the instructions for manufacturing proteins, among other things, and so inserting a functional gene into the body can, in theory, provide a lasting supply of whatever missing proteins a patient might need. But gene therapy has had a troubled history, in part because scientists cannot precisely control where a new gene inserts into a cell's DNA and how active it is once there (which determines how much protein is produced). These problems can lead to unwanted side effects–including the development of malignant tumors.

A logical solution to the problem of having proteins made in undesirable places and amounts would be to combine a therapeutic gene with a switch that could reliably turn it on or off as needed. As it happens, says Kirk–who is chairman and chief executive officer of Intrexon, a company that is developing new genetic engineering techniques–insects routinely use just such a switch to control molting.

Here's the thing. Insects do not just sort of molt, starting and then stopping partway; they either do it, or they don't. The genetic pathway that drives the process must remain completely turned off until the time is right. The gene that interests Kirk serves as the master switch for all this activity. It codes for a hormone called ecdysone. As ecdysone surges through the insect, it turns on a raft of other genes to start building the new exoskeleton. After the new exoskeleton is ready, the insect discards the old one. Once molting is nearing completion, the levels of ecdysone fall to zero—at which point the genetic pathway turns off. More important, from Intrexon's point of view, the switch is airtight when turned off—molting does not happen in the absence of ecdysone. The switch does not allow this group of genes to act together again until molting is set to begin.

The scientists at the company decided to take advantage of this foolproof characteristic to tightly control any genes transplanted into people. Imagine equipping each replacement gene with a biological switch that turns on—and thus activates the therapeutic gene—only in the presence of ecdysone molecules that have been adapted to work with human physiology. Patients given low doses of this activating drug (technically known as a ligand) could turn on only a few copies of the new gene, thus producing low amounts of whatever compounds were encoded. Patients given high doses of this activating drug could turn on many copies of the gene and thus manufacture large quantities of the related compound. In an unexpected emergency, however, withdrawing the ligand would shut the entire process down. No ecdysone, no activity by the introduced gene. As an added bonus, the ecdysone would not interfere with any other genes, which biologists call "cross talk," because the human body does not otherwise use, or need, the hormone to regulate genetic activity. Or, as Kirk puts it, "Humans don't molt."

Over the past eight years Intrexon has "wired" its ecdysone switches to thousands of human genes, demonstrating in laboratory tests that virtually any gene in the body can be put under the hormone's control. In addition, Kirk's group has added a second layer of checks by stitching in so-called cell-specific promoters to their switches. Cell-specific promoters are swatches of genetic material that cause genes to turn on (be "expressed") only in specific tissues, such as neurons or blood or liver cells, or only in certain conditions, such as the low-oxygen environment of a tumor. The addition of these molecular gatekeepers further reduces the chances of unwanted side effects occurring in parts of the body that had not been targeted for treatment.

Meanwhile other groups are borrowing from different biological processes to develop their own genetic switches and added control mechanisms. Eventually the ability to deliver several switch-controlled genes—each one able to be dialed up or down as needed—should make gene therapy safe and effective enough that it can become part of mainstream medicine. Or at least that is the idea. Preliminary tests in humans suggest that the switch approach could work as intended. So far it has been studied mostly in cancer and will likely make its biggest mark there first.

First Trials

In particular, investigators are studying the switch approach as a way of making cancer immunotherapy less of a harrowing ordeal for patients. Cancer immunotherapies, which have made a lot of headlines of late, aim either to reawaken an immune response that has been lulled to sleep by chemical signals from a malignancy or to jump-start an entirely new and more powerful anticancer response than a patient's immune system can achieve on its own. Trouble is, a rebooted immune system can easily slip into overdrive, triggering life-threatening fevers and the potentially lethal buildup of fluids throughout the body.

Gene switches are now being evaluated, for example, in small trials of carefully chosen patients with recurrent melanoma (a type of skin cancer) and breast cancer. Doctors inject one or two tumors in these individuals with genes designed to boost production of cytokines—signaling molecules (such as interferon and various interleukins) that the immune system uses to monitor and adjust the fight against tumors. Investigators believe that they may not need to treat all the malignant lesions in each patient, because once the immune system is properly primed to tackle one nest of cancerous cells, it should start searching for others elsewhere in the body without the need for further prompting.

Cytokines trigger a wide range of physiological reactions–from opening up blood vessels so that immune cells can rush to the scene of an infection to activating ruthless killer T cells, which, among other things, specialize in destroying cancerous cells. But to date, doctors have not been able to treat patients successfully with one of the most powerful cytokines, known as interleukin-12, or IL-12.

This failure stems in part from IL-12's propensity for unleashing a "cytokine storm," in which the immune system seemingly goes on a rampage against the body. In the bloodstream, IL-12 can cause a sharp drop in blood pressure, difficulties with lung function, and heart problems, which together can easily lead to organ failure and death. And yet, says Laurence Cooper, a physician-scientist at the University of Texas M.D. Anderson Cancer Center and CEO of biotech company Ziopharm Oncology, "there is a ton of literature on its effectiveness in the tumor microenvironment. IL-12 is the holy grail of immunotherapy." The idea, then, is to deliver as much IL-12 as possible to a single tumor but not so much that a cytokine storm occurs. Here is where the switch technology could prove revolutionary.

Researchers insert the switch-enabled IL-12 genes into an individual's tumor, where they take up residence in many cells, including the immune cells that are already there, giving the latter a boost. Because the switch can be activated only by the presence of the ligand, physicians can increase the levels of cytokine in the tumor very deliberately by slowly increasing the amount of drug they give their patient. If a cytokine storm starts to develop, they can skip the next scheduled dose, thereby averting the worst of the damage.

Ziopharm, which is working with Intrexon to develop switch-enabled cytokine treatment in people, reports encouraging results so far. Kirk acknowledges that they might have chosen to test their approach on something less potent than the IL-12 gene–where the slightest misstep could prove fatal. But he says, "We chose one of the hardest genes because we wanted to pressure-test the switch." In other words, when it counts most, does a switch that has been turned off remain completely off?

Results from two safety studies conducted at several medical centers (and totaling fewer than 40 patients) suggest that the answer is yes. Although no one was cured, the switch-controlled regimen appeared to be reasonably safe. As anticipated, a few patients did begin showing signs of a dangerous overreaction, but it dissipated soon after they stopped taking the ecdysone pills.

The researchers also found hints that the therapy could be helpful. In one of the studies, they injected the gene-plus-switch combination into 12 people with metastatic breast cancer. Each of them had already endured an average of eight previous cancer treatments, with diminishing hopes of survival. For various reasons, investigators were able to evaluate the new therapy's effect in only seven patients, however. The IL-12 treatment shrank some of their tumors, and in three people, the disease appeared to remain stable, at least for the short duration of the trial. The second safety study, in 26 patients who had been treated an average of six different times for metastatic melanoma, showed an uptick in cytokine levels and other cancer-fighting activity. In May 2015 Ziopharm initiated another study using switch-enabled IL-12 as an experimental treatment for glioblastoma multiforme, a particularly aggressive type of brain cancer.

Riboswitches

Richard Mulligan of Harvard Medical School has been working on a different kind of switch. His approach features small naturally occurring RNA molecules called ribozymes. First described in the 1980s, ribozymes are like enzymes in that they catalyze chemical reactions in the body, but most enzymes are proteins, and ribozymes consist of RNA. In a feature useful for switches, some ribozymes also have the ability to cut themselves up and induce genetic molecules to which they have been attached to self-destruct.

Mulligan's constructs give rise to a ribozyme linked not to a classic gene but to a messenger RNA (mRNA) molecule. When cells make proteins, they first copy the DNA in a gene into messenger RNA (a mobile, single-stranded transcript), after which the mRNA gets translated into the protein. From the cell's point of view, the addition of a stretch of DNA or its corresponding mRNA should result in the same outcome–production of a specific protein.

As a first step, researchers assemble and inject a strand of DNA that codes for a "self-cleaving" ribozyme plus the selected therapeutic protein. If a human cell transcribed this synthetic DNA into mRNA, the ribozyme portion would cut itself, causing the rest of the mRNA molecule to appear defective; the surrounding cellular machinery would then break the mRNA apart and cause the entire process of building a protein to grind to a halt. It would be as if the gene had been turned off.

Starting in 2000, Ronald R. Breaker and his colleagues at Yale University showed how to protect the mRNA but also to turn off protein synthesis when desired. The trick was to link the ribozyme to an additional molecule called an aptamer, which is a kind of sensor that is designed to be activated by a drug. In the presence of the drug (and only in the presence of the drug), the sensor changes shape in a way that prevents the ribozyme from destroying the mRNA. With the full length of mRNA intact, the cell makes the protein. When the drug that acts on the sensor is withdrawn, the ribozyme and the mRNA self-destruct.

By 2004 Mulligan and his colleagues were regularly equipping his ribozyme switches with carefully customized drug-sensitive sensors, and he continues to hone the technology today. The sensors can be designed with great specificity, Mulligan says, further reducing the chances of unwanted side effects. As with ecdysone, the mRNA that is connected to the riboswitch would allow production of the protein only if the patient swallowed the appropriate pill. Take the pill, and you have, in effect, turned on a gene. Stop the pill, and the gene stays off.

Multiple Switches

Although single gene switches are not yet perfected, investigators can envision a not too distant future in which multiple switches become the norm, allowing increasingly precise control of gene therapy. Combining switch-enabled gene therapy with other anticancer regimens may also bear tremendous fruit.

Already M.D. Anderson's Cooper, for example, is combining a couple of switch-enabled genes with a cell-based cancer treatment. The genes will contribute interleukin-12 and another cytokine, interleukin-15; lab tests suggest that IL-15 makes IL-12 even more effective at rallying immune cells. The third part of this experimental treatment–the cells–is a group of genetically engineered immune cells called CAR T cells that are better able to direct their firepower on cancerous tissue than naturally occurring immune cells can. Adding switch-bearing IL-12 and IL-15 genes to the CAR T cells should allow Cooper to boost the cells' potency and effectiveness. Because the gene switches and their respective activators will allow him to adjust the levels of IL-12 and IL-15 independently, he should be able to fine-tune the treatment to produce the best results with the least amount of IL-12, thereby further reducing the risk of unleashing a cytokine storm. With a touch of whimsy, Cooper calls this new suite of engineered cells "remote-control CARs."

Many technical hurdles must still be overcome, but the potential arc of progress is beginning to take shape. If inserting new genes into our bodies in the 1990s was Genetic Engineering 1.0, then the switch-based control of our genes is Genetic Engineering 2.0. Someday many of the pills that doctors give patients may be used to switch on various transferred genes at precisely the right place and time in the body instead of flooding every organ and tissue with the powerful pharmaceutical agents that act both where they are needed and elsewhere, causing side effects. Many drugs will no longer be manufactured in giant vats and so-called bioreactors in pharmaceutical facilities. Instead new gene treatments will allow patients to churn out a molecule exactly where and when it is needed most in the body.

Referenced

Semi-Synthetic Ecdysteroids as Gene-Switch Actuators: Synthesis, Structure–Activity Relationships, and Prospective ADME Properties. Silvia Lapenna et al. in *ChemMedChem*, Vol. 4, No. 1, pages 55–68; January 2009.

Small Self-Cleaving Ribozymes. Adrian R. Ferré-D'Amaré and William G. Scott in *Cold Spring Harbor Perspectives in Biology*, Vol. 2, No. 10, Article No. a003574; October 2010. http://cshperspectives.cshlp.org/content/2/10/a003574.long

The War on Cancer Is Evolving. Video featuring Laurence Cooper. October 15, 2013. www.acgtfoundation.org/video/dr-laurence-cooper-war-cancer-evolving

About the Author

Jim Kozubek is the author of Modern Prometheus: Editing the Human Genome with Crispr-Cas9*, published by Cambridge University Press.*

Mysterious 'Retron' DNA Helps Scientists Edit Human Genes

By Saugat Bolakhe

Deep in a bacterium's gelatinous matrix dwell little "cellular machines" called retrons, which produce single strands of DNA to detect certain viral infections. Now for the first time, researchers have used these natural DNA scriptwriters to modify genes in human cells. A new study, published in *Nature Chemical Biology*, suggests this technique can enhance gene editing across diverse animal groups.

Although the well-known CRISPR process has made gene editing much easier in recent years, it "has its own limitations," says the study's senior author Seth Shipman, a bioengineer at the University of California, San Francisco. This process introduces an enzyme called Cas9 to cut DNA segments and provides templates of desired DNA, designed by researchers, for cells to incorporate during the repair process. But this template DNA is created in the laboratory and must be inserted separately from CRISPR's components–and it does not always penetrate the cellular membrane.

Shipman and his colleagues instead used retrons to manufacture that DNA inside the cell itself, where the CRISPR process can readily use it. Retrons carry an enzyme called reverse transcriptase that builds DNA strands based on RNA. They also feature "some strangely overlapping loops of RNA" that help them function, says Santiago Lopez, a graduate student at U.C.S.F. and lead author on the study.

The researchers modified retrons in the lab so they would produce the desired template DNA. Additionally, they elongated the RNA loops, a change that turned out to let each retron produce more DNA copies. Finally, they inserted the retrons into cells along with CRISPR's components.

Using this process, retrons produced from 10 to 100 times more template DNA in yeast cells than in human cells. The retrons also achieved better editing precision in yeast than in human cells, possibly because of the differing number of strands or the way each cell type repairs DNA. "But frankly, we are not that worried right now," Shipman says, "since this is only a foot in the door." He says more adjustments and optimization will likely yield highly accurate editing in human cells.

"If we can repurpose retrons to produce DNA as 'donors' within a patient cell, it can be used for gene therapy applications for diseases such as sickle cell anemia, which require repair of only small stretches of faulty genetic sequences," says University of Nebraska molecular biologist Channabasavaiah B. Gurumurthy, who was not involved in the study.

But introducing foreign DNA into human tissue cells can also "elicit adverse immune responses that limit genetic modifications," says Jin-Soo Kim, director of South Korea's IBS Center for Genome Engineering, who was also not involved in the work. Researchers who are using CRISPR alone have developed processes to suppress such responses, Kim adds, but it remains to be seen how to accommodate retrons.

About the Author

Saugat Bolakhe is a Nepalese freelance science writer who covers the environment and life sciences.

Could the Next Big Information Technology Be ... DNA?

By James E. Dahlman

Billions of years before humans developed hard drives, evolution chose DNA to store its most precious information: the genetic code. Over time DNA became so proficient at this task that every known life-form on earth uses it. With recent technological breakthroughs that allow us to easily "read" and "write" DNA, scientists are now repurposing this age-old molecule to store new types of information—the kind that humans are generating at an exponential rate in the age of big data.

The concept of repurposing DNA to store information beyond genetic code has been discussed extensively. After all, the 1s and 0s of computer code are bumping up against the limits of physics. One of the challenges to safely storing all the data we create was exposed recently, when Myspace—once the most popular social network—announced that a decade's worth of data may have been irreparably lost in a server-migration project. The long-term protection of data, like those of a Web site that rebooted after a period of dormancy, exposes where existing technologies are vulnerable and clunky. And it's not just a spatial problem: significant energy is needed to maintain data storage.

The properties of DNA have the potential to get around these issues. For one thing, DNA's double-helix structure is perfectly suited for information storage because knowing the sequence of one strand automatically tells you the sequence of the other strand. DNA is also stable for extended periods, which means the integrity and accuracy of information can be maintained. For example, in 2017 scientists analyzed DNA isolated from human remains that were 8,100 years old. These remains were not even stored in ideal conditions the entire time. If kept in a cool, dry environment, DNA can almost certainly last tens of thousands of years. DNA is also stable for long stretches, which means the integrity and accuracy of information can be maintained.

Perhaps the most compelling aspect of the double helix, however, is that it can fold into an extraordinarily dense structure. For comparison, every individual human cell contains a nucleus with a diameter of approximately 0.00001 meter. Yet if the DNA inside a single nucleus was stretched out, it would reach two meters. Put another way, if the DNA in a person was strung together, it would extend 100 trillion meters. In 2014 scientists calculated that it is theoretically possible to store 455 exabytes of data in a single gram of DNA. This information-storage density is about a millionfold higher than the physical storage density in hard drives.

Although DNA has commonly been thought of as a storage medium, there are still significant scientific, economic and ethical hurdles to overcome before it might replace traditional hard drives. In the meantime, DNA is becoming more widely–and immediately–useful as a broader form of information technology. DNA has been used, for instance, to record old Hollywood films, preserving the classics in genetic code instead of fragile microfilm. Even more recently, DNA has been used as a tool to design safer gene therapies, speed up anticancer drug development and even generate what is perhaps the first genetic "live stream" of a living organism. On the frontiers of this evolving field, DNA is being pursued not just for long-term data storage but for facilitation of data generation at unprecedented speed. That is because DNA is more scalable than any other molecule in both directions: it allows us to dramatically expand the amount of data we create and shrink the resources needed to store them.

Accelerating New Nanoparticles

In recent years scientists have increasingly used DNA as a molecular recorder to understand and keep track of their experimental results. In many cases, this process involves DNA bar coding: To label and track the result of an individual experiment, scientists use a known DNA sequence to serve as a molecular tag. For example, one experimental outcome might be associated with the DNA sequence ACTATC, whereas another outcome might be associated with a TCTGAT, and so on.

DNA bar coding has been around since the early 1990s, when Richard Lerner and the late Sydney Brenner, both then at the Scripps Research Institute, proposed it as a way to track chemical reactions. Their concept was tremendously innovative but ahead of its time: technologies that easily and inexpensively read out DNA had not yet been developed. Its potential was only realized after many scientists made contributions to nucleotide chemistry, microfluidics and other approaches, which together enabled the advent of what is called next-generation sequencing. A major breakthrough came in 2005, when researchers reported that 25 million DNA bases were analyzed in a four-hour experiment.

Next-generation sequencing has continued to rapidly improve; it is now easy to read millions of DNA sequences at the same time, which means that thousands of experiments can be performed and analyzed simultaneously. Analyzing DNA bar code experiments with next-generation sequencing is its own form of data management: instead of testing ideas one at a time, scientists can make 20,000 predictions and test them all to see which is correct.

Biologists were the first to utilize DNA bar coding extensively. As it has become more accessible, researchers in many different fields, including chemical engineering and materials science, are using the technology to perform experiments at entirely new scales. In my laboratory at the Georgia Institute of Technology, for instance, engineers are using DNA bar codes to improve the design and function of nanoparticles so that they can safely deliver drugs to diseased cells. Nanotechnology, which relies primarily on physics and chemical engineering, may seem completely unrelated to DNA. But when you think of DNA as a way to track and store any data, its utility as an organizational tool becomes apparent.

One fundamental problem for nanotechnologists is that designing experiments to search for effective therapies is still far easier than performing them and analyzing the results. That is because the shape, size, charge, chemical composition and many other variables of individual nanoparticles can alter how well they deliver their genetic drugs to diseased cells. Additionally, these factors all interact with one another, making it a struggle for researchers to predict which nanoparticle will deliver its drug in the most targeted way. An obvious solution is to evaluate every nanoparticle one by one. But data from established pharmaceutical companies that have developed nanoparticles for RNA drugs have demonstrated that this type of testing can require several hundred million dollars to pull off.

That is where the storage capabilities of DNA can make big strides. To increase the number of nanoparticles we are able to test, we can design thousands of them with diverse chemical structures—large, positively charged spheres or small, neutrally charged triangles, for example—and assign each a DNA bar code.

Nanoparticle one, with chemical structure one, carries DNA bar code one. Nanoparticle two, with chemical structure two, carries DNA bar code two. We repeat this bar-coding process many times, thereby creating many different nanoparticles, each with its own unique molecular DNA tag. We can then administer hundreds of these nanoparticles to diseased cells. To identify the nanoparticle that most successfully delivered the drug, we use DNA sequencing to quantify the bar codes inside the cells.

The scale of such experiments is entirely new to nanomedicine. A "traditional" experiment in my field generates between one and five data points. By the end of 2019 my lab hopes to quantify how 500 different nanoparticles deliver gene therapies to 40 different cell types. Doing so is equivalent to running 20,000 experiments simultaneously.

As a result, we also needed to create a data-analysis pipeline capable of monitoring data quality, as well as helping us statistically test our results. First, we measured how well results from one replicated experiment predicted delivery in another. Once we knew the large data sets were reliable, we used statistics to ask whether certain nanoparticle traits—such as their size—affected delivery to target tissues. We found that the chemistry of the nanoparticle, not its size, dictated nanoparticle delivery. Using this approach, we hope to discover safe gene therapies more quickly, using far fewer resources. One of our goals is to identify a nanoparticle that can specifically deliver gene therapies that help kill tumors, thereby reducing side effects such as nausea and hair loss that accompany existing treatments.

We have already had some success. In 2018, by using very large data sets generated by DNA bar-coding experiments, we rapidly identified new nanoparticles that deliver gene therapies to endothelial cells, which line blood vessels, as well as several types of immune cells, which govern how our bodies respond to disease. This finding could change treatment by allowing us to change the activity of proteins in immune cells that are currently "undruggable," meaning the proteins are hard to target with small-molecule drugs or antibodies. As a result of data published in journals that included the *Proceedings of the National Academy of Sciences USA*, *Advanced Materials* and the *Journal of the American Chemical Society* in 2018 and 2019, we received a flood of interest from other gene therapists and were able to start GuideRx, a bar-coding company that focuses on efficiently developing safe gene therapies.

DNA bar coding has now become so commonplace that it is being applied in different ways even within a single field. One example is cancer biology, which looks at how genetic mutations cause cancer and how new drugs can treat it. Drug resistance remains a major challenge in this field: patients often initially respond to a drug but relapse as it loses the ability to kill tumor cells.

Scientists in the lab of Todd Golub at Harvard University have used DNA bar coding to study such resistance. In 2016 they described how they used a virus to permanently insert a DNA bar code directly into the genome of cancer cells. Cancer cell type A received bar code sequence A; cancer cell type B received bar code B, and so on. The scientists mixed the different cells together, plated them on a dish and treated them with a cancer drug.

If the drug killed the cancer cell or slowed its growth, then the cell would not divide. But if the cell became resistant to the drug, then it divided rapidly. Thus, over time the relative amount of bar code sequence A increased if cell type A became resistant to the drug or, alternatively, decreased if cell type A was killed by the drug. By sequencing all the bar codes from surviving cells over time, the lab quantified how well all the cell types responded to the drug simultaneously.

Later that year the lab of Monte Winslow at Stanford University used DNA-bar-coded pancreatic cell lines to identify drugs that prevented the spread of cancer, or metastasis. The lab bar coded each cell line using a virus, then plated each cell line in its own well. Each well was then treated with an anticancer drug. In this way, drug one became associated with bar code one. Immediately thereafter, the scientists injected the cells into the bloodstream, and they later measured which cells spread to the lungs. By identifying the bar codes that were abundant or absent, the researchers identified drugs that respectively promoted or prevented metastasis.

In a third example, scientists at the Broad Institute of the Massachusetts Institute of Technology and Harvard University used DNA bar coding to study how all the genes in the genome affect a single cancer. The researchers first grew a very large number of cells and plated them in a large dish together. Then they used a gene-editing system to inactivate or, alternatively, activate all the genes in the genome one by one. The sequence of the gene whose expression had been modulated acted as the bar code. By treating the cells with a cancer drug and sequencing the DNA over time, the scientists could understand how every gene in the genome affects drug resistance.

In these approaches, DNA is acting both as a data-generating molecule, because it is required to perform all the experiments simultaneously, and as a data-storage molecule, because next-generation sequencing is used to analyze the DNA bar codes. The implications are stunning: the same techniques can be applied to autoimmune and neurological diseases and cardiovascular dysfunction. The full power of using DNA bar coding can be understood with a simple exercise. In the examples discussed earlier, replace the word "cancer" with a different disease or the word "resistance" with any desired drug response. In this way, DNA bar coding is positioned to fundamentally streamline early-stage drug development, thereby accelerating the path to effective therapies.

Reading vs. Writing

DNA bar coding relies on "reading" known DNA sequences. Until recently, however, it was not practically possible to "write" DNA sequences. Broadly speaking, I think of writing DNA as purposefully converting other forms of information–such as pictures, movies or biological states–into sequences that can be stored and read out later. Many of these new writing technologies are driven by gene-editing systems derived from clustered regularly interspaced short palindromic repeats (CRISPR). With rationally engineered CRISPR systems, scientists can write DNA sequences.

Several of the most recent advances exploit the way CRISPR systems naturally evolved to defend bacteria against viral attacks. More specifically, viruses attack bacteria by binding onto the bacterial surface, then inserting their viral DNA or RNA. To "remember" the virus for future attacks, bacteria evolved CRISPR systems that identify viral DNA or RNA and then insert small snippets of the DNA into their own genome. In other words, the bacteria are "writing," or "recording," a history of the viruses that have attacked them to defend themselves.

By exploiting this mechanism, Seth Shipman, working in the lab of Harvard geneticist George Church and now at the University of California, San Francisco, used CRISPR to record images of a human hand directly into the genome of *Escherichia coli*. To accomplish this task, Shipman and his colleagues first expressed two proteins: Cas1 and Cas2. Together these proteins can acquire DNA nucleotides and insert them into the genome. The researchers then "fed" *E. coli* DNA sequences that encoded for pixels that–when sequenced together–created the image of a hand. Doing so required the scientists to assign different aspects of information to DNA. For example, in one case, A, C, G and T each stood for a different pixel color, whereas an associated DNA bar code sequence encoded the spatial position of the pixel within the entire image.

By sequencing the DNA from the *E. coli*, the authors then recapitulated the original image with more than 90 percent accuracy. Next, they repeated the experiment but with an important twist: they added the DNA at different times and included a method to analyze the position of the recorded DNA sequences, relative to one another. By measuring whether the sequences were added into the *E. coli* genome earlier or later, they were able to create a series of images, thereby encoding a movie. The researchers recorded a GIF from a part of the first motion picture, which was created by Eadweard Muybridge in 1878 and depicted a galloping horse. In a 2017 paper, they showed that they had reconstituted Muybridge's famous movie by sequencing the bacterial genome.

Even more recently, scientists in the lab of Randall Platt at the Swiss Federal Institute of Technology Zurich (ETH Zurich) made a critical discovery that takes these approaches even further by targeting mRNA, which is a key molecular cousin of DNA. Instead of recording images encoded by unnatural DNA sequences, they used a CRISPR system from a different bacterial species to generate so-called living records of natural mRNA gene expression in bacteria. The combination of all the different mRNAs in a cell dictates which proteins are made and therefore all cellular function.

To record mRNA produced by a cell at different time points, scientists at Platt's lab first screened CRISPR-Cas proteins derived from many different bacterial strains. This process allowed them to identify proteins capable of converting natural mRNA into DNA and encoding it into the genome. They found that Cas1 and Cas2 proteins from the bacterium *Fusicatenibacter saccharivorans* were capable of doing so. Through a series of elegant studies using specialized viruses, the team demonstrated in 2018 that the cells accurately recorded whether they had been previously exposed to oxidative stress, acidic conditions or even an herbicide.

These results were extremely exciting because they demonstrated that the genes naturally expressed by a cell at a given time could be recorded into the genome for later analysis. As Platt's lab continues to improve this technology, it is increasingly feasible that cellular recording could become commonplace. This development would enable scientists to track how a cell has become cancerous, responds to infection over time and even ages.

The Ubiquity of DNA Storage

As DNA is used to generate, track and store information in an increasing number of fields, the most obvious question is whether DNA will eventually compete with conventional electronic storage devices to maintain all the digital data humans generate. Currently the answer is no–hard drives and flash memory devices are far better at keeping information than even the most advanced DNA systems.

But like all technologies, conventional electronic devices have limitations. They take up physical space and require specific environmental conditions; even the most durable ones are unlikely to survive more than a few decades. Given these issues, it may soon become hard to maintain all the data we are generating today.

DNA, by comparison, could almost certainly last tens of thousands of years if kept in cool, dry conditions. It is already routinely stored at −20 or even −80 degrees Celsius in labs that require very cold conditions and can also be stored in the kind of extreme heat that typical electronics cannot withstand. In 2015 Robert Grass and Wendelin Stark, both at ETH Zurich, showed that DNA stored in silica could withstand 70 degree C temperatures for a week without introducing any errors. And although hard drives can fit as much as one terabit per square inch, recent estimations suggest that all the information generated in the entire world could theoretically be held in less than a kilogram of DNA.

There are still significant technological advances that need to be overcome for DNA storage to become commonplace. The primary limitation is that storing information is not identical to extracting it. Getting data from a hard drive is nearly instantaneous; extracting them from DNA requires sequencing, which currently takes a few minutes to a day to complete. And despite huge leaps in DNA sequencers over the past few years, they remain large and expensive as compared with hard drives.

These barriers are not the only considerations we must tackle before DNA storage can reach its full potential. As a society, we need to acknowledge that the ubiquity of DNA sequencing will also mean that it will become even easier to track people while generating new vulnerabilities for data security. Examples of privacy concerns abound, both in the U.S. and globally.

DNA sequencing is already being used by police departments across the U.S. with little oversight. By asking people who are under arrest–even for minor crimes–for their DNA, the police are establishing large data banks of genetic information. Some have argued this is the 21st-century equivalent of old-fashioned fingerprinting, but there is a critical difference. Fingerprints identify a single individual; if one of your relatives provides his or her DNA, that person is releasing information that can identify you or anyone else in your family. In China, under the guise of a health program, officials have gathered genetic information from nearly 36 million people. This population includes many Uighurs–members of a Muslim ethnic group that experiences discrimination. It remains unclear how these data will be used by the government.

Currently these concerns around DNA storage involve a person's genetic code itself–the discussion has been around protecting identity. But in the future, if other categories of information such as health care data, legal contracts and individual digital histories were stored in DNA, this scenario would launch even more questions about the vulnerability of DNA storage in the realms of both physical security and cybersecurity. Because so much information can be held in such a tiny space, how will data be distributed to avoid too much concentration in a single place? And even if extraction can be streamlined, how will data be routinely accessed and returned without exposing them to malicious hacks or accidental loss?

When I consider all the hard work–both scientific and ethical–that needs to be accomplished, it can seem daunting. I like to think about the Wright brothers because I grew up in the same Ohio town they did. Their first flight lasted 12 seconds and 37 meters. Sixty-six years later, without the advantages of modern computing, humans landed on the moon. These feats make me optimistic that we can harness the natural power of DNA over the next few decades and, by actively acknowledging its capability to do harm, help to ensure it mostly does good.

Referenced

Next-Generation Digital Information Storage in DNA. George M. Church et al. in *Science*, Vol. 337, page 1628; September 28, 2012.

High-Throughput In Vivo Screen of Functional mRNA Delivery Identifies Nanoparticles for Endothelial Cell Gene Editing. Cory D. Sago et al. in *Proceedings of the National Academy of Sciences USA*, Vol. 115, No. 42, pages E9944–E9952; October 16, 2018.

Transcriptional Recording by CRISPR Spacer Acquisition from RNA. Florian Schmidt et al. in *Nature*, Vol. 562, pages 380–385; October 18, 2018.

About the Author

James E. Dahlman is an assistant professor at the Wallace H. Coulter Department of Biomedical Engineering at the Georgia Institute of Technology and Emory University. His laboratory works at the interface of drug delivery, nanotechnology, genomics and gene editing.

Section 2: Fighting Disease

CRISPR Gene Editing May Help Scale Up Coronavirus Testing

By Jim Daley

Testing is one of the most daunting obstacles to overcome before thousands can again pack beaches and baseball stadiums. The much-lauded gene-editing technology CRISPR is now making a bid to help fill in holes in testing regimens. Last week researchers published a study in *Nature Biotechnology* describing a new assay for the novel coronavirus that causes COVID-19 that uses the technique to deliver results in about 40 minutes. The work began when study co-author Charles Chiu, an infectious disease physician at the University of California, San Francisco, was researching a CRISPR-based Lyme disease test earlier this year. Then SARS-CoV-2 began its fateful journey around the globe, and he quickly shifted the research in his laboratory.

To develop the probe, Chiu and his colleagues at U.C.S.F. collaborated with researchers at Mammoth Biosciences, which was co-founded by biochemist and CRISPR co-discoverer Jennifer Doudna. The test uses different reagents than the PCR-based SARS-CoV-2 ones that are currently in use, offering a potential alternative where there are shortages of the chemicals needed to conduct the latter assays. One drawback, however, is that the new approach's sensitivity, or ability to correctly provide positive results, is slightly lower than that of existing tests. Chiu says that it will take about two weeks to develop the CRISPR test for clinical lab use—and that a point-of-care version could be ready in as little as two to three months. *Scientific American* spoke with him about the technique.

[An edited transcript of the interview follows.]

Q: How does the test work?

A: CRISPR is a technology that allows you to target any particular gene, and you can think of it almost as a molecular scissor. It can very quickly and very precisely identify a gene, or basically a portion of DNA sequence, corresponding to a virus. This is why it's commonly thought of as a gene-editing tool. But the same technology that allows us to edit genes also allows you to detect them. In this case, we're targeting the genes of SARS-CoV-2. We use a protein called CRISPR-Cas12. If it finds the right target, it cuts that target. And by doing so, it releases a signal.

Q: The test detects the E (envelope) and N (nucleoprotein) genes on SARS-CoV-2. Why did you choose to target these?

A: There are a couple of reasons. In particular, the N gene is the gene that is most highly expressed in viral infections. In other words, it's the most common, so it's the easiest to detect. The second reason is that the N gene is effectively the same target that the Centers for Disease Control and Prevention's PCR test targets. And this was the first test the Food and Drug Administration cleared for emergency use authorization (EUA), so we're simply leveraging the fact that there's already a test targeting that gene, which happens to be highly expressed. We selected the E gene because it's one of the targets used by the World Health Organization. Our goal was not to reinvent the wheel but rather to use the same targets that have already been shown to be effective in detecting SARS-CoV-2.

Q: Where does this test fit into the existing landscape of coronavirus testing—and PCR-based tests in particular?

A: We're nowhere near the scale of testing that we really need in order to be confident that we can reemerge and go back to work. Although there are many different PCR tests, there's still a need to drastically ramp up both the scale and the speed of testing. And every test that's based on PCR uses a common set of reagents and, as a result, is subject to shortages in their availability. Our assay uses enzymes and reagents that are different than nearly all of the other tests out there, so it's an alternative that could really ramp up testing. We also need technology that allows us to do testing in point-of-care settings: in the emergency room, in the doctor's office or even at home. CRISPR technology provides a pathway by which we could eventually do that. It doesn't require bulky instrumentation. And the reagents themselves are quite cheap.

Q: Do you think CRISPR-based testing will replace PCR assays?

A: I don't. It is slightly less sensitive than the PCR tests that are now commonly available. CRISPR technology still has a long way to go before we can achieve comparable sensitivity–although in principle, there's no theoretical reason we can't. But I do think that when the time comes, and it's more widely adopted and optimized, it's certainly possible, because CRISPR-based tests are faster and potentially cheaper. And in this particular test, we might be able to bypass the extraction step of converting a sample into RNA or DNA. So there's a chance it might be able to replace conventional testing. But it's still a new technology, and I don't think that's going to happen anytime soon. In the immediate term, it's meant to be a complimentary tool.

Q: Your approach has a 95 percent sensitivity rate. Is that good enough for mass testing?

A: I think it is. Patients who are infected with SARS-CoV-2 usually have the highest viral load when they initially develop symptoms. And in fact, it may even be prior to symptom onset that they may have very high viral loads. In hospitalized patients, the viral loads fluctuate over time, but generally, they go down. So when patients are minimally symptomatic, that's when they're incredibly infectious, and we need to find a way to break the chain of transmission. The promise of CRISPR is that it can provide a way to have cheap, decentralized testing for a large fraction of the population. Right now the need is not to detect every low-level infection, it's to get testing implemented at a widespread scale.

Q: Some PCR-based tests can provide results in under 15 minutes. This test takes about 40 minutes. Can you make it any faster?

A: In the interest of getting the test out there and maximizing its sensitivity, we chose to make it more robust. So the steps are actually longer than they need to be. Speeding up the test would involve bypassing the extraction step, as well as decreasing the reaction times. We've been able to decrease the detection time down to 10 minutes, so it's definitely doable. The question is, Can you do that and still retain the same sensitivity?

Q: Is it an inexpensive test?

A: Overall the test is quite inexpensive, because the reagents and enzymes can be made in bulk. The cost right now is going to be in terms of trying to manufacture the right platform. If we can package the test into a cheap, disposable cartridge, it potentially could be very cheap. The cost of reagents is just a few dollars per test. For the clinical-lab-based CRISPR test, we're aiming for less than $1 per test, and the cost of a home-based test is under $5. The key, though, is whether [we] can manufacture [it] at high volume and high capacity and keep it cheap–and still be able to run the test in different settings outside of the laboratory.

Q: What are the technical issues associated with scaling up testing?

A: The availability of reagents is one. Another simply involves the need for centralized testing. Typically, you need a molecular lab to run these instruments, so testing is limited by the number of instruments you can deploy. The Abbott ID NOW [assay], for example, is one of the few point-of-care instruments. But it runs only one sample at a time. We need to be able to run hundreds of tests at a time. And CRISPR, in addition to allowing you to do point of care because the reaction is very fast and simple, [might let you] potentially run thousands of samples a time in a clinical laboratory–and do so faster than PCR can. You can also potentially distribute it for home use. Multiplexing the

CRISPR test in a point-of-care setting or at home is feasible if the test consists of a disposable cartridge containing all of the reagents you need.

Q: Are there other places using CRISPR for infectious disease testing?

A: Yes, there's a group at the Broad Institute [of the Massachusetts Institute of Technology and Harvard University] developing CRISPR-Cas13 enzymes for SARS-CoV-2 testing. And because it's useful for diagnostics, as well as gene editing, we envision developing CRISPR to be both a diagnostic tool and a therapy that could not only detect the virus but destroy it.

Q: What potential applications are there for using this test on emerging viruses in the future?

A: What's great about this technique is that these probes can be designed on the fly, which means we can rapidly change our target. I was actually collaborating with Mammoth Biosciences on this test to develop a tick-borne Lyme disease test. Once the SARS-CoV-2 genome sequence was available, we went from having the sequence to having a functional test in two or three weeks. Eventually, we envision a point-of-care test that would be programmable for any target you wish.

About the Author

Jim Daley is a freelance journalist from Chicago. He writes about science and health.

How Designer DNA Is Changing Medicine

By Carolyn Barber

For as long as he could remember, Razel Colón had known pain. It ripped down his neck and back, shot through his legs and traveled on to his feet, often leaving him writhing and incapacitated. He suffered occasional attacks of "acute chest," in which breathing suddenly becomes difficult. "It felt like an elephant was sitting on my chest, with tight, tight pain," Colón tells me. Trips to the emergency department and the hospital were commonplace. "If I was lucky," he says, "I could stay away for a month."

Colón, from Hoboken, N.J., is just 19, but the sickle cell disease that produced these effects had been a constant, if unwelcome, companion. But he tells his story now from the perspective of one who has gone a year and a half without that pain. He can do things that previously were out of the question: play basketball, lift weights, swim in cold water. His treatment, says his long-time physician Stacey Rifkin-Zenenberg, a pediatric hematologist-oncologist at Hackensack University Medical Center, "changed him from having the disease to being a carrier."

Colón's case represents a point on the curve of an emerging technology that may forever alter our approach to treating diseases like sickle cell. That world, the cutting-edge world of innovative genomic therapies, is once again in the midst of explosive change–and designer DNA lies at the heart of the conversation.

This is daring new territory. Some strategies, such as gene therapy, have been available for some time, including the ability to genetically modify cells in order to produce a therapeutic effect–that is, to add a corrected gene into the genome in order to try to treat disease. Traditionally, viruses have been used to deliver healthy genes into cells, but the past decade has been witness to profound change. Multiple gene therapies have been approved to treat a variety of conditions: squamous cell skin cancer, a rare form of inherited blindness, melanoma, blood disorders and so on.

It was this type of treatment, as part of the largest lentivirus ongoing gene therapy trial led by Bluebird Bio, that effectively arrested Colón's sickle cell condition. Unpublished interim data from 19 participants in the trial followed for at least six months, with a history of severe vaso-occlusive events (VOEs), or sickle cell crises similar to Colón's, demonstrated complete resolution of severe VOEs in all patients, according to a company spokesperson. The trial is ongoing, and data are not yet complete, so caution is prudent, but "the promise is tremendous," the spokesperson said.

The next-generation technology, gene editing, is another level altogether. Gene editing enables scientists to precisely target abnormal genes of many organisms (bacteria, plants, animals), snip the DNA, then remove, replace or add new DNA at the incision site. "Imagine you have a car with a flat tire," says Fyodor Urnov, a gene editing expert at the University of California, Berkeley, and the Innovative Genomics Institute. "Gene therapy is taking a fifth wheel and putting it somewhere on the car and hoping it runs. Gene editing is repairing the flat."

The technology received a huge boost with the arrival in 2012 of a gene-editing tool called CRISPR, an acronym for Clustered Regularly Interspaced Short Palindromic Repeats. The CRISPR technology is easier to use, cheaper and more efficient than older genome editing methods, enabling scientists to quickly alter DNA sequences to modify gene function. That could positively affect the health of the organism, or even reverse disease symptoms.

"It is often described as 'molecular scissors,'" says Jennifer Doudna, its co-inventor and Nobel laureate in chemistry. "Scientists can harness CRISPR to not just cut specific locations in DNA of any organism, but also to provide a template to repair the DNA."

In sickle cell disease, or SCD, a single mutation in the beta-hemoglobin gene leads to red blood cells that become crescent-shaped, or sickled. These sickled cells are sticky, and they clog arteries, preventing adequate oxygen delivery to tissues in the body. This can cause acute debilitating pain episodes, such as the kinds that Colón experienced, and may result in any number of complications: anemia, strokes and organ damage involving the lungs, heart, kidney, spleen, etc.

Patients often have a poor quality of life because of repeated hospitalizations and transfusions, and face the prospect of early deaths. In regions like Africa and the Middle East where health care resources are much more limited, many children die of SCD before their fifth year of life.

CRISPR is accelerating the pace at which treating such disease via genetic engineering is moving. Matthew Porteus, a gene-editing pioneer, founder of CRISPR Therapeutics and professor of pediatrics at Stanford School of Medicine, says researchers currently employ two primary gene-editing strategies in their attempt to cure sickle cell patients. One type uses CRISPR to essentially flip a genomic switch, turning on healthy fetal hemoglobin production again, which was shut down early in life. The advantage? The fetal hemoglobin doesn't sickle.

A second gene editing strategy, gene correction, directly fixes the mutation in a faulty gene that has caused disease. In the case of sickle cell, the correction allows the body simply to produce normal hemoglobin. Researchers have logged a staggering amount of work trying to get to this point.

While progress has been made with the approval of several new drugs to help alleviate the symptoms of SCD, they are not curative. Bone marrow (stem cell) transplants are the only option for cure, but finding healthy donor matches can be challenging. Enter the genomic therapies, which Theodore Friedmann of the University of California, San Diego, first proposed for genetic disease in 1972, and which took a step toward reality the same year with the production of recombinant DNA by Paul Berg at Stanford University. In the 1980s, scientists showed how DNA could be delivered into cells, and by 2003 the entire human genome had been deciphered.

There were setbacks, and even when gene editing systems came to be as the new century opened, using them remained challenging and time-consuming. Then, came CRISPR, slicing and dicing its way into the clinical arena, poised to potentially change not only treatment but prevention of disease. The technology is already being employed by scientists researching cancer, lymphoma, AIDS, cystic fibrosis and more, for diagnostics including SARS-CoV-2 detection, and even in agricultural efforts to engineer larger tomatoes, non-browning apples, and longer-lasting mushrooms.

In real time, the hope is that genetic editing will provide a cure for a very common and debilitating inherited disease, sickle cell. Several experts with whom I spoke suggested, for the first time, that a cure for SCD may be in the offing.

That is an astounding thought. In practice, the ability to excise nonworking genes and replace them with normally functioning ones may help blunt the worst effects of a wide variety of diseases. When the 2020 Nobel Prize in Chemistry was awarded to Jennifer Doudna and Emmanuelle Charpentier for their invention of this technology, the secretary-general of the Royal Swedish Academy of Sciences, Goran K. Hansson, put it plainly: "This year's prize is about rewriting the code of life."

And scientists are already working on next-gen CRISPR technology that's more precise. "Base editors," which correct single-letter DNA mutations without cutting the DNA double helix, have been shown recently to treat sickle cell disease in mice. And then there are "prime editors" which can replace even larger DNA snafus.

It's all pretty remarkable–and it's still early. "The field of genomic therapies for the hemoglobinopathies is not a zero-sum game," says Urnov. "I am convinced ultimately that there will be multiple approved medicines. There will be multiple gene therapy approvals, and there will be multiple gene-editing approvals."

There remain many questions to consider, including the moral implications of how far a concept like gene editing might be taken, as well as cost, safety and accessibility (experts say current treatments can run to $2 million). Says Doudna, "A true cure means a treatment for everyone who needs it, which is why we're hard at work on the next generation of therapies to bring down the cost and make it more accessible." And, as gene editing is not perfect, "The long-term safety of all the genetic modification therapies will have to be studied carefully," says Porteus.

In the coming months, a University of California Consortium (U.C. San Francisco, U.C. Berkeley and U.C.L.A.), will conduct the first human open label phase I–II trial using nonvirally delivered CRISPR technology developed at the Innovative Genomics Institute. Scientists hope to alleviate the suffering of sickle cell trial participants by substituting their faulty beta-globin gene–the one that causes the disease–with a corrected one, directly fixing the mutation in their blood stem cells.

"The more of the mutations that are corrected and propagated in healthy red blood cells, the more likely a cure is to follow," says Mark Walters, the trial's principal investigator and a professor of pediatrics at U.C. San Francisco. "Sickle cell disease will be cured using genomic therapies." That is the highest use of the technology, after all–and it may become a reality even as we continue to learn about the field.

This is an opinion and analysis article; the views expressed by the author or authors are not necessarily those of Scientific American.

About the Author

Carolyn Barber has been an emergency department physician for 25 years. She is co-founder of the homeless work program Wheels of Change and author of many articles and the book Runaway Medicine: What You Don't Know May Kill You, *which was recently Amazon's number-one Hot New Release in Health Care Administration. She received her M.D. at the Johns Hopkins University School of Medicine.*

The Definition of Gene Therapy Has Changed

By Esther Landhuis

Three decades after its first, faltering steps in humans, gene therapy is emerging as a treatment option for a small but growing number of diseases. Although the concept faced scientific and ethical uncertainty when it was floated in the 1970s, the foundation of the approach—replacing or fixing a single, disease-causing gene—has proved solid. Researchers have developed different ways to correct or influence the way someone's genes function and used those techniques to create therapies for several blood disorders, as well as degenerative eye and muscle diseases. More than half a dozen such treatments have gained approval in the U.S. in the past five years, and numerous others, aimed at a variety of conditions, are progressing toward clinical trials.

Existing gene therapies rely on two fundamental approaches. The more common approach draws blood from the patient and reprograms specific cells within the laboratory before reinjecting them into the person's body. The other method delivers gene treatments directly into the body, usually to easier-to-reach areas such as the eye. Now the field is beginning to mature and move beyond these initial tactics. Continued advances have made gene delivery safer and more effective, leading to dozens of human trials in new tissues, such as the liver and heart. Other approaches are pushing beyond the original definition of gene therapy, with cutting-edge molecular tools that fix errors within genes rather than replacing or inserting a whole gene.

Yet despite recent progress, gene therapy faces numerous hurdles on the path to wider clinical use—chief among them is how to target specific tissues without triggering an immune response. Broader, long-term challenges include improving both manufacturing efficiency and cost: Gene therapy treatments in the U.S. currently average more than $400,000 per dose. Nevertheless, with so much potential and so many patients in need of new solutions, gene therapy will only continue to grow in both prominence and potency.

Early Success and Shock Waves

The concept underlying the original gene therapy approaches, some of which are still in use, is fairly straightforward: When a disease results from a missing or dysfunctional gene, deliver a functional copy of the gene into affected cells. That, says Prashant Mali, a bioengineer at the University of California, San Diego, was the "version 1.0 definition of gene therapy."

One of the first attempts came in 1990, when researchers at the National Institutes of Health treated two young girls with severe immunodeficiency caused by a missing enzyme. In that trial, as with many current treatments, the therapeutic genes needed to produce the enzyme hitched a ride into the target cells inside engineered viruses, which had large chunks of their genome stripped out. This rendered the virus unable to replicate while making space for the delivery of the needed human genes. In essence, says Charles Gersbach, director of the Center for Advanced Genomic Technologies at Duke University, the approach capitalized on the virus's ability to infect human cells while "taking advantage of the viral shell as a Trojan horse to deliver therapeutic gene cargo."

The NIH team drew some of the girls' blood to isolate white blood cells, which were then "infected" with the viruses that carried the gene encoding the missing enzyme. Next the team infused the corrected cells into the girls. Each child received about a dozen more infusions over the next 18 to 24 months. The treatment wasn't a cure, but it lessened their symptoms and proved the approach could be used safely. That, in and of itself, was "a major milestone," Gersbach says.

A flurry of new gene therapy trials quickly followed, but in 1999 18-year-old Jesse Gelsinger died when an experimental gene treatment designed to treat his metabolic liver disease sent his immune system into overdrive. A few years later, in 2003, researchers reported that several people treated for immunodeficiency developed leukemias, an unfortunate result of the virus randomly inserting its cargo into cancer-promoting regions of the genome.

Researchers began to think, "'Wait a minute, maybe we don't understand this as well as we thought we did,'" Gersbach says. Gene therapy stalled for the better part of a decade. Clinical trials on hold, researchers turned all their attention back to the lab—studying and tweaking viral vectors, removing additional genes, and treating them with chemicals to make them safer and more effective at reaching target cells.

The renewed focus provided time and space for a better understanding of what worked and what didn't. Today, because of that progress, many gene therapies employ adeno-associated virus (AAV) or retrovirus vectors, each with their own pros and cons, in addition to improved versions of the adenovirus vector from the earliest trials. The genetic cargo delivered by most AAV vectors remains within the cell as separate, free-floating elements rather than stably integrating into the host cell's genome. That makes the vectors far less likely than earlier vectors to induce cancer but can make a treatment less durable, depending on how long the therapeutic genes remain in the host cell. On the other hand, because they are small, they can infect a broad range of cells and spread efficiently within tissues. Retroviruses offer different advantages. They can hold larger and more complex genes than AAV vectors. And some, such as lentiviruses, tend to insert themselves into coding regions, the parts of the genome that get translated into proteins. This minimizes cancer risk while conferring longer-lasting benefits than AAV vectors.

A Field Reawakens

Gene therapy got a fresh start in the early 2010s, when researchers in Pennsylvania and Maryland independently reported results from trials for the treatment of leukemia or lymphoma. The experimental therapies trained and turbocharged the patients' immune systems so they could detect and destroy cancer cells. To do this, the scientists had to engineer genes that would equip cells to recognize and kill tumors. They put those genes into retroviral vectors and delivered them to T cells, immune cells that had been isolated from the subjects' blood. When the treated T cells were reinfused, they put the cancer into remission. "Everything was looking really promising again," says Cynthia Dunbar, a physician-scientist who studies blood cell treatments at the National Heart, Lung, and Blood Institute.

The U.S. Food and Drug Administration has since approved several of these T cell treatments, known as chimeric antigen receptor (CAR) T cell therapies, for certain lymphomas and leukemias, as well as multiple myeloma. Because CAR T cell treatments don't address gene dysfunction per se but rather endow T cells with tumor-hunting capabilities, some have debated whether they qualify as gene therapies at all. Methodologically, though, CAR T boosts cell function by using viral vectors to deliver genes–similar to the earliest forays. "What you define as 'gene therapy' is a little bit gray on the edges," Dunbar says.

Another genetic approach that is a bit gray around the categorical edges is known as oligonucleotide therapy. Rather than correcting existing genes, this technique uses short sequences of nucleic acids, or oligonucleotides, to influence how cells translate genes into proteins. One such treatment, nusinersen (Spinraza), binds to intermediary RNA molecules to trick cells into making more of a protein that is missing in people with spinal muscular atrophy.

Gene Therapy 2.0

In the past decade technological advances have ushered in a new era, and the definition of gene therapy continues to evolve, Mali says. The newest approaches forgo the delivery of healthy genes and instead aim to precisely repair the gene within the cell. When there is a mutation or other error in the genome, Mali says, now the question is, "Could we actually go in and fix it?"

This innovation is fueled by the Nobel Prize–winning discovery of CRISPR-Cas9, an immune defense system in bacteria that detects specific DNA sequences of invading viruses and directs an enzyme to slice up and destroy the viral genome. The system has utility far beyond bacteria: Scientists found they could also use it to make precise cuts within the mammalian genome. In just seven years the technique has moved from in vitrolab experiments in mammalian cells to human trials.

The "cargo" in CRISPR-based therapies is not a piece of DNA but the gene-editing system itself, introduced into cells either by a virus, within a nanoparticle, or on its own as an RNA-protein complex. The therapies can be used ex vivo (outside the body) to alter cells in the lab before returning them to the patient or by sending gene-editing tools directly to affected tissues, where they edit cellular genomes.

Several dozen companies are now developing such CRISPR-based therapies. One early-phase clinical trial employed an ex vivo gene-editing method to treat people with sickle cell disease or with a related blood disorder called beta thalassemia. Those researchers reported results for their first two subjects in January 2021. And in June another company reported the first-ever successful trial of direct gene editing, which used nanoparticles to deliver CRISPR-Cas9 components into liver cells and inactivate a gene implicated in a rare disease called transthyretin amyloidosis.

Emerging methods have allowed for greater precision and nuance—exchanging individual nucleotides, for example, or temporarily dampening a gene's activity without changing its DNA—giving researchers the latitude to set their sights ever higher. They are working on treatments for neurological diseases, autoimmune disorders and additional cancers. Over the long term they aim to move beyond single-gene disorders to treat conditions caused by interactions of multiple genes, such as cardiovascular disease and chronic pain. With gene treatments successfully alleviating some illnesses, researchers, clinicians and patients are hoping to sustain the progress of the past decade and establish gene therapy as a cornerstone of modern medicine.

About the Author

Esther Landhuis is a freelance science and health journalist based in the San Francisco Bay Area.

Four Success Stories in Gene Therapy

By Jim Daley

After numerous setbacks at the turn of the century, gene therapy is treating diseases ranging from neuromuscular disorders to cancer to blindness. The success is often qualified, however. Some of these therapies have proved effective at alleviating disease but come with a high price tag and other accessibility issues: Even when people know that a protocol exists for their disease and even if they can afford it or have an insurance company that will cover the cost—which can range from $400,000 to $2 million—they may not be able to travel to the few academic centers that offer it. Other therapies alleviate symptoms but don't eliminate the underlying cause.

"Completely curing patients is obviously going to be a huge success, but it's not [yet] an achievable aim in a lot of situations," says Julie Crudele, a neurologist and gene therapy researcher at the University of Washington. Still, even limited advances pave the way for ongoing progress, she adds, pointing to research in her patients who have Duchenne muscular dystrophy: "In most of these clinical trials, we learn important things."

Thanks to that new knowledge and steadfast investigations, gene therapy researchers can now point to a growing list of successful gene therapies. Here are four of the most promising.

Gene Swaps to Prevent Vision Loss

Some babies are born with severe vision loss caused by retinal diseases that once led inevitably to total blindness. Today some of them can benefit from a gene therapy created by wife-and-husband team Jean Bennett and Albert Maguire, who are now ophthalmologists at the University of Pennsylvania.

When the pair first began researching retinal disease in 1991, none of the genes now known to cause vision loss and blindness had been identified. In 1993 researchers identified one potential target gene, *RPE65*. Seven years later Bennett and Maguire tested a therapy targeting that gene in three dogs with severe vision loss–it restored vision for all three.

In humans, the inherited condition that best corresponds with the dogs' vision loss is Leber congenital amaurosis (LCA). LCA prevents the retina, a layer of light-sensitive cells at the back of the eye, from properly reacting or sending signals to the brain when a photon strikes it. The condition can cause uncontrolled shaking of the eye (nystagmus), prevents pupils from responding to light and typically results in total blindness by age 40. Researchers have linked the disease to mutations or deletions in any one of 27 genes associated with retinal development and function. Until gene therapy, there was no cure.

Mutations in *RPE65* are just one cause of inherited retinal dystrophy, but it was a cause that Bennett and Maguire could act on. The researchers used a harmless adeno-associated virus (AAV), which they programmed to find retinal cells and insert a healthy version of the gene, and injected it into a patient's eye directly underneath the retina. In 2017, after a series of clinical trials, the Food and Drug Administration approved voretigene neparvovecrzyl (marketed as Luxturna) for the treatment of any heritable retinal dystrophy caused by the mutated *RPE65* gene, including LCA type 2 and retinitis pigmentosa, another congenital eye disease that affects photoreceptors in the retina. Luxturna was the first FDA-approved in vivo gene therapy, which is delivered to target cells inside the body (previously approved ex vivo therapies deliver the genetic material to target cells in samples collected from the body, which are then reinjected).

Spark Therapeutics, the company that makes Luxturna, estimates that about 6,000 people worldwide and between 1,000 and 2,000 in the U.S. may be eligible for its treatment–few enough that Luxturna was granted "orphan drug" status, a designation that the FDA uses to incentivize development of treatments for rare diseases. That wasn't enough to bring the cost down. The therapy is priced at about $425,000 per injection, or nearly $1 million for both eyes.

Despite the cost, Maguire says, "I have not yet seen anybody in the U.S. who hasn't gotten access based on inability to pay."

Those treated show significant improvement: Patients who were once unable to see clearly had their vision restored, often very quickly. Some reported that, after the injections, they could see stars for the first time.

While it is unclear how long the effects will last, follow-up data published in 2017 showed that all 20 patients treated with Luxturna in a phase 3 trial had retained their improved vision three years later. Bennett says five-year follow-up with 29 patients, which is currently undergoing peer review, showed similarly successful results. "These people can now do things they never could have dreamed of doing, and they're more independent and enjoying life."

Training the Immune System to Fight Cancer

Gene therapy has made inroads against cancer, too. An approach known as chimeric antigen receptor (CAR) T cell therapy works by programming a patient's immune cells to recognize and target cells with cancerous mutations. Steven Rosenberg, chief of surgery at the National Cancer Institute, helped to develop the therapy and published the first successful results in a 2010 study for the treatment of lymphoma.

"That patient had massive amounts of disease in his chest and his belly, and he underwent a complete regression," Rosenberg says—a regression that has now lasted 11 years and counting.

CAR T cell therapy takes advantage of white blood cells, called T cells, that serve as the first line of defense against pathogens. The approach uses a patient's own T cells, which are removed and genetically altered so they can build receptors specific to cancer cells. Once infused back into the patient, the modified T cells, which now have the ability to recognize and attack cancerous cells, reproduce and remain on alert for future encounters.

In 2016 researchers at the University of Pennsylvania reported results from a CAR T cell treatment, called tisagenlecleucel, for acute lymphoblastic leukemia (ALL), one of the most common childhood cancers. In patients with ALL, mutations in the DNA of bone marrow cells cause them to produce massive quantities of lymphoblasts, or undeveloped white blood cells, which accumulate in the bloodstream. The disease progresses rapidly: adults face a low likelihood of cure, and fewer than half survive more than five years after diagnosis.

When directed against ALL, CAR T cells are ruthlessly efficient–a single modified T cell can kill as many as 100,000 lymphoblasts. In the University of Pennsylvania study, 29 out of 52 ALL patients treated with tisagenlecleucel went into sustained remission. Based on that study's results, the FDA approved the therapy (produced by Novartis as Kymriah) for treating ALL, and the following year the agency approved it for use against diffuse large B cell lymphoma. The one-time procedure costs upward of $475,000.

CAR T cell therapy is not without risk. It can cause severe side effects, including cytokine release syndrome (CRS), a dangerous inflammatory response that ranges from mild flulike symptoms in less severe cases to multiorgan failure and even death. CRS isn't specific to CAR T therapy: Researchers first observed it in the 1990s as a side effect of antibody therapies used in organ transplants. Today, with a combination of newer drugs and vigilance, doctors better understand how far they can push treatment without triggering CRS. Rosenberg says that "we know how to deal with side effects as soon as they occur, and serious illness and death from cytokine release syndrome have dropped drastically from the earliest days."

Through 2020, the remission rate among ALL patients treated with Kymriah was about 85 percent. More than half had no relapses after a year. Novartis plans to track outcomes of all patients who received the therapy for 15 years to better understand how long it remains effective.

Precision Editing for Blood Disorders

One new arrival to the gene therapy scene is being watched particularly closely: in vivo gene editing using a system called

CRISPR, which has become one of the most promising gene therapies since Jennifer Doudna and Emmanuelle Charpentier discovered it in 2012–a feat for which they shared the 2020 Nobel Prize in Chemistry. The first results from a small clinical trial aimed at treating sickle cell disease and a closely related disorder, called beta thalassemia, were published this past June.

Sickle cell disease affects millions of people worldwide and causes the production of crescent-shaped red blood cells that are stickier and more rigid than healthy cells, which can lead to anemia and life-threatening health crises. Beta thalassemia, which affects millions more, occurs when a different mutation causes someone's body to produce less hemoglobin, the iron-rich protein that allows red blood cells to carry oxygen. Bone marrow transplants may offer a cure for those who can find matching donors, but otherwise treatments for both consist primarily of blood transfusions and medications to treat associated complications.

Both sickle cell disease and beta thalassemia are caused by heritable, single-gene mutations, making them good candidates for gene-editing therapy. The method, CRISPR-Cas9, uses DNA sequences from bacteria (clustered regularly interspaced short palindromic repeats, or CRISPR) and a CRISPR-associated enzyme (Cas for short) to edit the patient's genome. The CRISPR sequences are transcribed onto RNA that locates and identifies DNA sequences to blame for a particular condition. When packaged together with Cas9, transcribed RNA locates the target sequence, and Cas9 snips it out of the DNA, thereby repairing or deactivating the problematic gene.

At a conference this past June, Vertex Pharmaceuticals and CRISPR Therapeutics announced unpublished results from a clinical trial of beta thalassemia and sickle cell patients treated with CTX001, a CRISPR-Cas9-based therapy. In both cases, the therapy does not shut off a target gene but instead delivers a gene that boosts production of healthy fetal hemoglobin–a gene normally turned off shortly after birth. Fifteen people with beta thalassemia were treated with CTX001; after three months or more, all 15 showed rapidly improved hemoglobin levels and no longer required blood transfusions. Seven people with severe sickle cell disease received the same treatment, all of whom showed increased levels of hemoglobin and reported at least three months without severe pain. More than a year later those improvements persisted in five subjects with beta thalassemia and two with sickle cell. The trial is ongoing, and patients are still being enrolled. A Vertex spokesperson says it hopes to enroll 45 patients in all and file for U.S. approval as early as 2022.

Derailing a Potentially Lethal Illness

Spinal muscular atrophy (SMA) is a neurodegenerative disease in which motor neurons–the nerves that control muscle movement and that connect the spinal cord to muscles and organs–degrade, malfunction and die. It is typically diagnosed in infants and toddlers. The underlying cause is a genetic mutation that inhibits production of a protein involved in building and maintaining those motor neurons.

The four types of SMA are ranked by severity and related to how much motor neuron protein a person's cells can still produce. In the most severe or type I cases, even the most basic functions, such as breathing, sitting and swallowing, prove extremely challenging. Infants diagnosed with type I SMA have historically had a 90 percent mortality rate by one year.

Adrian Krainer, a biochemist at Cold Spring Harbor Laboratory, first grew interested in SMA when he attended a National Institutes of Health workshop in 1999. At the time, Krainer was investigating how RNA mutations cause cancer and genetic diseases when they disrupt a process called splicing, and researchers suspected that a defect in the process might be at the root of SMA. When RNA is transcribed from the DNA template, it needs to be edited or "spliced" into messenger RNA (mRNA) before it can guide protein production. During that editing process, some sequences are cut out (introns), and those that remain (exons) are strung together.

Krainer realized that there were similarities between the defects associated with SMA and one of the mechanisms he had been studying–namely, a mistake that occurs when an important exon is inadvertently lost during RNA splicing. People with SMA were missing one of these crucial gene sequences, called *SMN1*.

"If we could figure out why this exon was being skipped and if we could find a solution for that, then presumably this could help all the [SMA] patients," Krainer says. The solution he and his colleagues hit on, antisense therapy, employs single strands of synthetic nucleotides to deliver genetic instructions directly to cells in the body. In SMA's case, the instructions induce a different motor neuron gene, *SMN2*, which normally produces small amounts of the missing motor neuron protein, to produce much more of it and effectively fill in for *SMN1*. The first clinical trial to test the approach began in 2010, and by 2016 the FDA approved nusinersen (marketed as Spinraza). Because the therapy does not incorporate itself into the genome, it must be administered every four months to maintain protein production. And it is staggeringly expensive: a single Spinraza treatment costs as much as $750,000 in the first year and $375,000 annually thereafter.

Since 2016, more than 10,000 people have been treated with it worldwide. Although Spinraza can't restore completely normal motor function (a single motor neuron gene just can't produce enough protein for that), it can help children with any of the four types of SMA live longer and more active lives. In many cases, Spinraza has improved patients' motor function, allowing even those with more severe cases to breathe, swallow and sit upright on their own. "The most striking results are in patients who are being treated very shortly after birth, when they have a genetic diagnosis through newborn screening," Krainer says. "Then, you can actually prevent the onset of the disease–for several years and hopefully forever."

About the Author

Jim Daley is a freelance journalist from Chicago. He writes about science and health.

Section 3: What's the Big Deal About GMOs?

New Gene-Editing Techniques Could Transform Food Crops–or Die on the Vine

By Stephen S. Hall

The hundred or so farmers crowding the ballroom of the Mendenhall Inn in Chester County, Pennsylvania, might not have had a background in gene editing, but they knew mushrooms. These local growers produce a staggering 1.1 million pounds of mushrooms on average every day, which is one reason Pennsylvania dominates the annual $1.2-billion U.S. market. Some of the mushrooms they produce, however, turn brown and decay on store shelves; if you've ever held a slimy, decomposing, formerly white mushroom in your hand, you know why no ones buys them. Mushrooms are so sensitive to physical insult that even careful "one-touch" picking and packing can activate an enzyme that hastens their decay.

On a foggy morning last fall, at a continuing education seminar on mushrooms, a biologist named Yinong Yang took the podium to deliver news of a possible solution for the browning problem. Yang, a cheerfully polite professor of plant pathology at Pennsylvania State University, is not an expert in the field. ("The only thing I know about mushrooms is how to eat them," he says.) But he edited the genome of *Agaricus bisporus*, the most popular dinner-table mushroom in the Western world, using a new tool called CRISPR.

The mushroom farmers in the audience had probably never heard of CRISPR, but they understood it was a big deal when Yang showed a picture of actress Cameron Diaz awarding inventors Jennifer Doudna and Emmanuelle Charpentier the Breakthrough Prize in November 2014, which came with a check for $3 million each. And they understood the enormous commercial implications when Yang showed them photographs comparing brown, decayed mushrooms with pristine white CRISPR-engineered *A. bisporus*, the all-purpose strain that annually accounts for more than 900 million pounds of white button, cremini and portobello mushrooms. (Penn State understood the commercial implications, too; the day before Yang's talk, the university filed for a patent on the mushroom work.)

In its brief three years as a science story, CRISPR has already generated more fascinating subplots than a Dickens novel. It is a revolutionary research tool with dramatic medical implications, thorny bioethical conundrums, an awkward patent spat and, floating over it all, billion-dollar commercial implications for medicine and agriculture. The technique has blown through the basic research community like an F5 tornado. Academic laboratories and biotech companies are chasing novel treatments for diseases such as sickle-cell anemia and beta-thalassemia. And there has even been speculation about DIY artists and bioentrepreneurs creating everything from purple-furred bunnies to living, breathing gene-edited tchotchkes, like the miniaturized pigs recently made in China as pets. The prospect of using CRISPR to repair embryos or permanently edit our DNA (a process known as human germ-line modification) has sparked fevered talk of "improving" the human species and calls for international moratoriums.

The CRISPR revolution may be having its most profound–and least publicized–effect in agriculture. By the fall of 2015 about 50 scientific papers had been published reporting uses of CRISPR in gene-edited plants, and there are preliminary signs that the U.S. Department of Agriculture, one of the agencies that assesses genetically modified agricultural products, does not think all gene-edited crops require the same regulatory attention as "traditional" genetically modified organisms, or GMOs. With that regulatory door even slightly ajar, companies are racing to get gene-edited crops into the fields and, ultimately, into the food supply.

The transformative aspect of CRISPR lies in its unprecedented precision. CRISPR allows you to knock out any gene or, with a little more effort, to add a desirable trait by inserting a gene in a specific place in a genome. This makes it, according to its practitioners, the least biologically disruptive form of plant breeding that humans have ever devised–including the "natural" breeding techniques that have been practiced for thousands of years. It also enables scientists to sidestep, in many cases, the controversial techniques of inserting DNA from other species into plants; these "transgenic" crops, such as the Monsanto-made corn and soybeans that are resistant to the herbicide Roundup, have aroused particular ire in GMO critics and led to public distrust of the technology. Yet some scientists are optimistic that CRISPR crops are so fundamentally different that they will change the tenor of the debate over GMO foods. "The new technology," says Daniel F. Voytas, an academic and company-affiliated scientist, "is necessitating a rethinking of what a GMO is."

Will consumers agree? Or will they see CRISPR crops as the latest iteration of Frankenfood–a genetic distortion of nature in which foreign (and agribusiness-friendly) DNA is muscled into a species, with unpredictable health or environmental consequences? Because CRISPR is only now being applied to food crops, the question has not yet surfaced for the public, but it will soon. Farmers such as Yang's mushroom growers will be the first to weigh in–probably in the next year or two.

Moments after Yang's talk, an industry scientist confronted him with the central challenge of CRISPR food. The researcher conceded Yang's point that the improved mushrooms required minimal tinkering with DNA compared with conventional GMOs. "But," the scientist said, "it is genetic modification, and some people will see it as we are playing God. How do we get around that?"

How well Yang and other scientists applying these gene-editing techniques to food can answer that question will determine whether CRISPR is a potentially transformational tool or one that will be stymied by public opposition.

"Wow, That's the One!"

The telltale sign of any transformational technology is how quickly researchers apply it to their own scientific problems. By that standard, CRISPR ranks among the most powerful additions to biology's tool kit in the past half a century. The gene-edited mushroom is a case in point.

Yinong Yang—his first name means "also practices agriculture" in Chinese—never worked with mushrooms until 2013, but you might say he was to the task bred. Born in Huangyan, a city south of Shanghai known as the citrus capital of China, he dabbled with some primitive gene-editing enzymes in the mid-1990s as a graduate student at the University of Florida and later at the University of Arkansas. He vividly remembers opening the August 17, 2012, issue of *Science*, which contained a paper from Doudna's lab at the University of California, Berkeley, and Charpentier's lab describing CRISPR's gene-editing potential. "Wow," he thought. "That's the one!" Within days he was hatching plans to improve traits in rice and potato plants through gene editing. His lab published its first CRISPR paper in the summer of 2013.

He was not alone. Plant scientists jumped on CRISPR as soon as the technique was published. Chinese scientists, who quickly embraced the technology, shocked the agricultural community in 2014 when they showed how CRISPR could be used to make bread wheat resistant to a long-standing scourge, powdery mildew.

The gene-editing revolution had begun before the arrival of CRISPR, however. For people like Voytas, CRISPR is merely the latest chapter in a much longer scientific saga that is just now bearing fruit. He first attempted gene editing in plants 15 years ago, while at Iowa State University, with a technology known as zinc fingers; his first gene-editing company foundered on patent issues. In 2008 he moved to the University of Minnesota and in 2010 patented, with former Iowa State colleague Adam Bogdanove, now at Cornell University, a gene-editing system in plants based on TALENs, a subsequent gene-editing tool. That same year Voytas and his colleagues started a company now known as Calyxt. Without the hoopla of CRISPR, agricultural scientists have used TALENs to produce gene-edited plants that have already been grown in fields in North and South America. Calyxt, for example, has created two strains of soybean modified to produce a healthier oil, with levels of monounsaturated fats comparable to olive and canola oils.* And the company has gene-edited a potato strain to prevent the accumulation of certain sugars during cold storage, reducing the bitter taste associated with storage, as well as the amount of acrylamide, a suspected carcinogen, produced when potatoes are fried.

Because these genetic modifications did not involve the introduction of any foreign genes, the USDA's Animal and Plant Health Inspection Service (APHIS) decided last year that the crops do not need to be regulated as GMOs. "The USDA has given regulatory clearance to plant a potato variety and two soybean varieties, so the potato and one of the soybean varieties are in the field this year," Voytas told me last October. "They basically considered these as just standard plants, as if they were generated by chemical mutagens or gamma rays or some nonregulated technology. The fact that we got regulatory clearance and can go almost immediately from the greenhouse to the field is a big plus. It allows us to really accelerate product development."

Animal scientists have also jumped on the gene-editing bandwagon. Researchers at the small Minnesota-based biotech firm Recombinetics have genetically blocked the biological signal that governs the growth of horns in Holstein cattle, the workhorse of the dairy industry. Ag scientists tout this application of gene editing as a more humane form of farming because it spares male and female Holstein cattle from a gruesome procedure during which dairy farmers physically gouge out and then cauterize developing horns (the procedure is done to protect both dairy cattle and dairy farmers from injury).**

Scott Fahrenkrug, the company's CEO, says the process does not involve transgenes, just the introduction of a few letters of DNA "to match the food we already eat." Korean and Chinese scientists, meanwhile, have teamed up to produce a pig with much more muscle mass, by using gene editing to knock out a gene called myostatin.

The speed, ease and thrift of CRISPR make it an even more attractive technique than TALENs. "Without a doubt," Voytas says, in the future CRISPR "is going to be the plant-editing tool of choice."

But the murky patent situation—both the University of California and the Broad Institute (run jointly by the Massachusetts Institute of Technology and Harvard University) claim to have invented CRISPR—may slow commercial agricultural development. DuPont recently reached a "strategic alliance" with Caribou Biosciences, a biotech associated with U.C. Berkeley, to use CRISPR applications in agriculture, but executives at two small biotechs told *Scientific American* that they were wary of developing CRISPR-related products while the patent dispute remains unresolved.

That's not a big issue for academic labs. The mushroom story took a decisive turn in October 2013, when a Penn State alum named David Carroll popped into Yang's lab. Carroll, who happened to be president of Giorgi Mushroom, wondered if new gene-editing techniques could be used to improve mushrooms. Emboldened by the power of CRISPR to create highly precise mutations, Yang replied, "What kind of trait do you want?" Carroll suggested antibrowning, and Yang immediately agreed to try it.

Yang knew exactly which gene he wanted to target. Biologists had previously identified a family of six genes, each of which encode an enzyme that causes browning (the same class of genes also triggers browning in apples and potatoes, both of which have been targeted by gene editors). Four of the so-called browning genes churn out that enzyme in the fruiting body of mushrooms, and Yang thought that if he could shut down one of them through a gene-editing mutation, he might slow the rate of browning.

The brilliant ease of CRISPR derives from the fact that it is straightforward for biologists to customize a molecular tool—a "construct"—that creates such mutations. Like a utility knife that combines a compass, scissors and vise, these tools excel at two tasks: homing in on a very specific stretch of DNA and then cutting it (the vise, or scaffolding, holds everything in place during the cutting). The homing is accomplished by a small piece of nucleic acid called the guide RNA, which is designed to mirror the DNA sequence in the target area and attach to it using the unique and specific attraction of DNA base pairs made famous by James Watson and Francis Crick (where As grab onto Ts and Cs grab onto Gs). If you make a piece of guide RNA that is 20 letters long, it will find its mirror sequence of DNA—with GPS-like precision—amid the string of 30 million letters that spell out the *Agaricus* mushroom genome. The cutting is then accomplished by the Cas9 enzyme, originally isolated from bacterial cultures in yogurt, which rides in on the back of the guide RNA. (The term "CRISPR/Cas9" is a bit of a misnomer now because CRISPR refers to clustered regularly interspersed short palindromic repeats, patches of DNA that occur only in bacteria. It is the Cas9 protein, loaded with an RNA targeting sequence, that edits plant, fungal and human DNA, even though no CRISPRs are involved.)

Once gene editors cut DNA at the desired spot, they let nature perform the dirty work of mutation. Any time the double helix of DNA is cut, the cell notices the wound and sets out to repair the break. These repairs are not perfect, however, which is exactly what makes CRISPR so powerful at creating mutations. During the repair process, a few letters of DNA usually get deleted; because a cell's protein-making machinery reads DNA in three-letter "words," deleting a couple of letters subverts the entire text and essentially inactivates the gene by creating what is known as a reading frame shift. That is precisely what happened with the gene-edited mushroom. In Yang's work, a tiny deletion of DNA inactivated one of the enzymes that promote browning—a mutation that Yang and his colleagues confirmed with DNA analysis. Editing complete. According to Yang, a skillful molecular biologist could in about three days build a custom-designed mutation tool to edit virtually any gene in virtually any organism.

That sentiment echoes the mantra scientists constantly invoke about CRISPR: it is fast, cheap and easy. It took about two months of lab work to create the antibrowning mushroom; Yang's demeanor suggested that the work was routine, if not ridiculously easy. And it was remarkably inexpensive. The trickiest step, making the guide RNA and its scaffolding, cost a couple of hundred dollars; a number of small biotech firms now make custom-order CRISPR constructs to edit any gene desired. The biggest cost is manpower: Xiangling Shen, a postdoctoral fellow in Yang's lab, worked on the project part-time. "If you don't consider manpower, it probably cost less than $10,000," Yang says. In the world of agricultural biotech, that is chump change.

And that doesn't begin to hint at the potentially game-changing thrift of CRISPR in the regulatory arena. Last October, Yang gave an informal presentation of the mushroom work to federal regulators at the USDA's APHIS, which decides if genetically modified food crops fall under government regulatory control (in short, whether they are considered a GMO); he came away from the meeting convinced that USDA regulators did not believe the CRISPR mushroom would require special or extended regulatory review. If true, that may be the most important way CRISPR is cheaper: Voytas has estimated that the regulatory review process can cost up to $35 million and take up to five and a half years.

Another advantage of the mushroom as a proof of principle for CRISPR in agriculture is the speed at which fungi grow: from spawn to maturity, mushrooms take about five weeks, and they can be grown year-round in windowless, climate-controlled facilities known as mushroom houses. The gene-edited soybeans and potatoes created by Calyxt, in contrast, take months to field-test, which is why the company sought, and received, regulatory clearance to grow its soybeans in Argentina last winter (2014–2015). "You bop back and forth over the equator," Voytas says, "so you can get multiple plantings in a year." Calyxt harvested its first North American gene-edited crops from the field last October.

One of the long-standing fears about genetic modification is the specter of unintended consequences. In the world of biotech foods, this usually means unexpected toxins or allergens making modified foods unhealthy (a fear that has never been documented in a GMO food) or a genetically modified crop running amok and devastating the local ecology. CRISPR is even making people such as John Pecchia think about unintended economic consequences. One of two mushroom professors at Penn State, Pecchia spends a lot of time in a low-slung cinder-block building situated on the outskirts of the campus, which houses the only center of academic mushroom research in the U.S. In the spring of 2015 Pecchia took some of Yang's starter culture and grew up the first batch of gene-edited mushrooms. Standing outside a room where a steamy, fetid mix of mushroom compost was brewing at 80 degrees Celsius, he notes that a mushroom with a longer shelf life might result in smaller demand from stores and also enable unexpected competition. "You could open up the borders to foreign mushroom imports," he adds, "so it's a double-edged sword."

In the tortuous path of genetically modified foods to market, here is one more paradox to chew on. No one knows what the gene-edited mushroom tastes like. They've been steamed and boiled, but not for eating purposes. Every mushroom created so far has been destroyed after Yang conducted browning tests. Once proof of principle has been established, Pecchia says, "we just steam them away."

Transgene-Free Modification

Will the public steam, sauté or otherwise welcome gene-edited food into their kitchens and onto their plates? That may be the central question in the most intriguing chapter in the CRISPR food story, which coincides with a crucial juncture in the tumultuous, 30-year debate over genetically modified crops.

When Yang described his mushroom project to the Pennsylvania farmers—and to officials at the USDA last October—he used a telltale phrase to describe his procedure: "transgene-free genetic modification." The phrase is a carefully crafted attempt to distinguish the new, high-precision gene-editing techniques like CRISPR from earlier agricultural biotech, where foreign DNA (transgenes) were added to a plant species. For Yang and many others, that delicate wording is important in recasting the GMO debate. Indeed, the acronym "GEO" (for gene-edited organism) has begun to crop up as an alternative to "GMO" or "GM."

The reframing is as much philosophical as semantic, and it is unfolding as the Obama administration is overhauling the system by which the government reviews genetically modified crops and foods. Known as the Coordinated Framework for Regulation of Biotechnology, this regulatory process, which has not been updated since 1992, defines roles for the USDA, the Food and Drug Administration, and the Environmental Protection Agency. The power of CRISPR has added urgency to the regulatory rethink, and scientists are using the opportunity to revisit a very old question: What exactly does "genetically modified" mean? Voytas, whose track record of publications and patents in gene-edited food crops makes him a sort of editor in chief of small agricultural biotechs in the U.S., answered with a grim little laugh when asked that question: "The GM term is a tricky one."

What's so tricky about it? Most critics of biotech food argue that any form of genetic modification is just that, genetic modification, bringing with it the possibility of unintended mutations or alterations that could pose risks to human health or the environment. Scientists such as Voytas and Yang reply that *all* forms of plant breeding, dating all the way back to the creation of bread wheat by Neolithic farmers 3,000 years ago, involve genetic modification and that the use of traditional breeding techniques is not a biologically benign process. It creates, as Yang put it, "huge" genetic disruptions. (Nina Fedoroff, a plant biologist and former president of the American Association for the Advancement of Science, has referred to domesticated versions of bread wheat, created by traditional breeding, as "genetic monstrosities.")

Before the era of recombinant DNA in the 1970s, which allowed first-generation agricultural biotech, plant breeders typically resorted to brute-force methods (x-rays, gamma rays or powerful chemicals) to alter the DNA of plants. Despite this blunderbuss approach, some of these random, man-made mutations modified genes in a way that produced desirable agricultural traits: higher yields, or more shapely fruit, or an ability to grow in adverse conditions such as drought. These beneficial mutations could then be combined with beneficial traits in other strains but only by crossing—or mating—the plants. That type of crossbreeding takes a lot of time (often five to 10 years), but at least it is "natural."

But it is also very disruptive. Any time DNA from two different individuals comes together during reproduction, whether in humans or plants, the DNA gets scrambled in a process known as chromosomal reassortment. Spontaneous mutations can occur in each generation, and millions of base pairs of DNA can be transferred when breeders select for a desired trait. It is natural, yes, but also "a big mash-up," according to Voytas. "In that process, you don't just move one gene," he says. "You often move a pretty big chunk of DNA from the wild species." Moreover, the desirable trait often drags along with it an undesirable trait on the same piece of DNA during the process of breeding; this "linkage drag" can actually harm the naturally bred plant. On the basis of several recent findings on the genetics of rice plants, some biologists hypothesize that domestication has inadvertently introduced "silent" detrimental mutations as well as obvious beneficial traits.

Although CRISPR is more precise than traditional breeding, the technique is not infallible. The precision cutting tool sometimes cuts an unintended region, and the frequency of these "off-target" cuts has raised safety concerns (it is also the main reason that gene editing of human sperm and egg cells is still considered unsafe and unethical). Jennifer Kuzma, a policy analyst at North Carolina State University, who has followed the science—and politics—of GMO agriculture since its inception, says, "That precision has merit, but it doesn't necessarily correlate with risk reduction," adding that off-target cuts "may introduce a different pathway to hazard." Feng Zhang of the Broad Institute (which holds the patent that is now being disputed) has published several refinements in the CRISPR system that improve specificity and reduce off-target hits.

The ease and relative thrift of CRISPR have also allowed academic labs and small biotechs back into a game that has historically been monopolized by big agribusinesses. Only deep-pocketed companies could afford to run the costly regulatory gauntlet in the beginning, and to date, almost every crop modification created by genetic engineering was done to enhance the economics of food production for farmers or companies, be it the increased yields of Monsanto's herbicide-resistant field crops or the shipping hardiness of Calgene's ill-fated Flavr-Savr tomato. Those genetic crop modifications were more appealing to agribusiness than consumers, and they were not very food-centric. As a group of agricultural policy experts at the University of California, Davis, recently observed, "the multinational corporations that have dominated the field for the past decade and a half do not have a glowing record in terms of innovation beyond traits for pesticide and herbicide resistance."

The new players have brought a different kind of innovation to agriculture. Voytas, for example, argues that the precision of gene editing is allowing biotech scientists to target consumers by creating healthier, safer foods. Voytas and his colleague Caixia Gao of the Chinese Academy of Sciences have pointed out that plants have many "antinutritionals": noxious self-defense substances or outright toxins that could be gene-edited away to improve nutritional and taste traits. Calyxt's gene-edited potato, for example, reduces a bitter taste trait associated with cold storage of the tubers.

But Voytas goes even further. He believes the Calyxt soybean could be sold to farmers as a non-GMO product because, unlike 90 percent of soybeans grown in the U.S., the gene-edited strains do not have any transgenes. "A lot of people don't want GM products," he says. "We could maybe make non-GM soybean oil and non-GM soybean meal with our product."

Like any powerful new technology, CRISPR has inspired some agricultural dreamers to envision almost science-fiction scenarios for the future of farming–scenarios that are already making their way into the scientific literature. Michael Palmgren, a plant biologist at the University of Copenhagen, has proposed that scientists can use the new gene-editing techniques to "rewild" food plants, that is, to resurrect traits that have been lost during generations of agricultural breeding. A number of economically significant food crops–notably rice, wheat, oranges and bananas–are highly susceptible to plant pathogens; the restoration of lost genes could increase disease resistance. The idea, Palmgren and his Danish colleagues recently noted, aspires to "the reversal of the unintended results of breeding."

Attempts at rewilding are already under way but with a twist. Rather than restoring lost wild traits to domestic breeds, Voytas says his University of Minnesota lab is attempting what he calls "molecular domestication": transferring agriculturally desirable genes from existing hybrids back into wild species that are hardier and more adaptable, such as the ancestral form of corn, and potatoes. "It's usually only a handful of critical changes that occurred–five, six or seven genes–that allowed a weedy species to become desirable, such as changes in fruit size or corn ear number, those sorts of things," Voytas says. Rather than crossing the wild varieties with the domesticated strains, which would require a 10-year breeding regime, he says, "maybe we can just go in and treat those genes and domesticate the wild variety."

There are early signs that gene editing, including CRISPR, may also enjoy a speedier regulatory path. So far U.S. regulators appear to view at least some gene-edited crops as different from transgenic GMO crops. When Calyxt first asked the USDA if its gene-edited potatoes required regulatory review, federal officials took about a year before concluding, in August 2014, that gene editing did not require special consideration; when the company went back to the USDA last summer with its gene-edited soybeans, government reviewers took only two months to reach a similar conclusion. To companies, this suggests that U.S. authorities view the new techniques as fundamentally distinct from transgenic methods; to critics, it suggests a regulatory loophole that companies are exploiting. Yang's mushrooms may be the first CRISPR food considered by the USDA.

And new technologies like CRISPR are forcing some governments to reconsider the definition of a GMO. Last November the Swedish Board of Agriculture decreed that some plant mutations induced by CRISPR do not meet the European Union's definition of a GMO, and Argentina has similarly concluded that certain gene-edited plants fall outside its GMO regulations. The E.U., which has historically restricted genetically modified plants, is currently reviewing policy in light of the new gene-editing techniques, but its oft-delayed legal analysis will not be made public until the end of March at the earliest. While there is not much middle ground, Voytas and others have suggested one potential compromise: gene editing that causes a mutation, or "knock out," should be viewed as analogous to traditional forms of plant breeding (where x-rays, for example, are used to create mutations), whereas gene editing that introduces new DNA (a "knock in") deserves regulatory scrutiny on a case-by-case basis.

The day of food-market reckoning for gene-edited crops may not be too far off; Voytas estimates that Calyxt will have a "small commercial launch" of its soybeans by 2017 or 2018. "It's going to take some time to get enough seed for, say, half a million acres," he says. "But we're pushing as hard and fast as we can."

How will the public respond? Kuzma predicts that people who have historically opposed genetic modification will not be drinking CRISPR Kool-Aid anytime soon. "The public that opposed first-generation GMOs is not likely to embrace this second generation of genetic engineering, just because you're tweaking a little bit of DNA," she says. "They're just going to lump it together with GMOs." Kuzma is more concerned about the need to revamp the overall regulatory structure and bring more voices into the review process, at an "inflection point" at which more and more gene-edited foods are wending their way to the marketplace.

And what about the mushroom? Beyond polite applause at the end of Yinong Yang's talk, the reaction of mushroom farmers remains unclear. Yang acknowledged as much when he told the farmers, "Whether this can be commercialized, that's up to you guys." For now, the antibrowning mushroom is just a lab project, a proof of principle. If growers are unconvinced of the value of the antibrowning mushroom or fear consumers will shun it, the well-edited mushroom may never see the light of day. That's usually a good thing for a mushroom, which grows in the dark, but is perhaps more ominous for a new and potentially transformative technology.

**Editor's Note (4/12/16): This sentence from the print article was edited after it was posted online. The original misstated the type fats comparable to olive and canola oils.*

***Editor's Note (5/10/16): This paragraph from the print article was edited after it was posted online. The original incorrectly stated that only the cows, rather than both the male and female Holstein cattle, were involved in the gene editing.*

Referenced

Precision Genome Engineering and Agriculture: Opportunities and Regulatory Challenges. Daniel F. Voytas and Caixia Gao in *PLOS Biology*, Vol. 12, No. 6, Article No. e1001877; June 10, 2014.

Conflicting Futures: Environmental Regulation of Plant Targeted Genetic Modification. Adam Kokotovich and Jennifer Kuzma in *Bulletin of Science, Technology & Society*, Vol. 34, Nos. 3–4, pages 108–120; June–August 2014.

Feasibility of New Breeding Techniques for Organic Farming. Martin Marchman Andersen et al. in *Trends in Plant Science*, Vol. 20, No. 7, pages 426–434; July 2015.

A Face-Lift for Biotech Rules Begins. Emily Waltz in *Nature Biotechnology*, Vol. 33, No. 12, pages 1221–1222; December 2015.

Tweaking Vegetables' Genes Could Make Them Tastier—And You'll Get to Try Them Soon

By Meghan Bartels

Today's Brussels sprouts taste better than you might remember from childhood. It's not that your refined adult palate appreciates them better. Rather a new variety has displaced the original vegetable. You can thank plant breeders for the change. And modern breeders, armed with new gene-editing technology, are looking to replicate Brussels sprouts' reinvention.

In the late 1990s scientists identified specific chemicals, called glucosinolates, that made Brussels sprouts taste bitter. Plant breeders started growing old seeds, previously discarded for producing paltry harvests, to identify tastier versions with lower levels of these compounds. Then they crossed these delicious but low-yield plants with modern, more prolific individuals until they found a descendant that made plenty of tasty sprouts, transforming the once maligned vegetable from a bitter pill into a popular side dish.

But other veggies haven't fared as well. That's because most breeding decisions favor plant traits that matter to vegetable growers, not vegetable eaters. "I'd say disease resistance is probably the major focus these days of most breeding programs because that's what imperils the ability of the farmer to grow the crop," says Harry Klee, a professor emeritus of horticulture at the University of Florida, who specializes in tomatoes. "Quality traits are really completely ignored."

In addition, breeders who focus on consumer crops must navigate the controversy surrounding genetically modified organisms (GMOs). Although all domesticated species now have different genetics than their ancestors, in agriculture, the term GMO refers to a plant that carries genes imported from a completely different species–and these changes are subject to stricter regulation. But newer techniques are allowing breeders to work within the context of a plant's own genome, making tweaks that don't trigger complicated rules.

Flavor is challenging to target because different people have different preferences–plus, even under the best conditions, flavor quality is more complex than something like yield. "We've spent a lot of time and money figuring out what flavor is, and most breeding programs don't have the capacity to measure those things," Klee says.

But interest in prioritizing flavor is starting to build, thanks in part to new genetic technology such as the gene-snipping technique CRISPR and DNA sequencing that is cheap enough to use liberally. "There's never been a better time to be a fruit breeder or a vegetable breeder because we have more tools and techniques," says Susan Brown, an apple breeder at Cornell University.

Some companies are beginning to use those tools to tackle the challenge of developing tastier veggies. One company, Pairwise, is fighting the same compounds that plagued Brussels sprouts: glucosinolates. But this time researchers are modifying salad greens–and they're armed with the science of gene editing.

Although kale, for instance, is particularly healthy, many prefer eating the less bitter romaine or iceberg lettuce. So Pairwise scientists figured out how to use CRISPR to edit a kalelike mustard green to fit that palate. They wanted to turn off the genes that code for an enzyme called myrosinase, which breaks down the glucosinolates and creates bitterness once the leaf is chewed in a diner's mouth. The result is a healthy but less bitter green that the company is marketing this year under the brand Conscious Foods.

This is an example of where flavor-minded gene editing can shine, says Tom Adams, co-founder and CEO of Pairwise. "From a gene-editing perspective, I think where taste comes in is that we can remove things that people don't like," he says. "It's a lot more difficult to think about how you can bring in the really complex, great tastes." To create those more complex tastes, Adams says, traditional breeding is still the best path.

Traditional breeding is the cornerstone of another high-tech flavor effort as well, one that seeks to reverse the ways of thinking that got us to tasteless vegetables in the first place. Instead of growing varieties that can withstand the storage and transportation needs of the agricultural system, a company called Plenty is shrinking the vast distance from the field to the table. Plenty grows its plants in indoor vertical farming facilities closer to consumers, so produce stays fresher, says Nate Storey, the company's co-founder and chief science officer.

When Plenty decided to start with greens, he says, the team grew thousands of traditionally bred varieties in its facilities. Then the researchers adopted only the ones that resulted in the tastiest crops rather than trying to develop new varieties. "There's no need to rebuild a wheel," Storey says of the technique. "We just screen all the wheels that exist and find the ones that work best in our system."

This method does not always succeed, however. The company couldn't find a tomato that thrived in its facilities, so it's working on developing its own variety using a sped-up version of traditional breeding.

Tomatoes are a popular target. A third company is creating more flavorful tomatoes using epigenetics–changing the expression of genes instead of the genes themselves. Whereas Pairwise is snipping out the gene sequences that produce an enzyme that interferes with flavor, Sound Agriculture is programming gene expression. This approach dials down production of undesirable compounds by making their genetic sequences less accessible for transcription.

Understanding how precisely to modify expression to get a desired result is still a work in progress, says Travis Bayer, co-founder and chief technology officer of Sound Agriculture. "The science of epigenetics in plants is really exciting and it's something that is evolving pretty rapidly," he says. The company's first epigenetically grown product, a tomato dubbed Summer Swell, is due to hit shelves this spring. Other projects in the works focus on leafy greens, as well as a handful of fruits, Bayer says.

All these vegetable growers hope that more flavorful products on store shelves will convince people to consume the recommended allowances of fruits and vegetables–and do so better than decades of nutritional guidance have.

"Don't waste your time talking about trying to educate people to eat better," says Klee, the tomato breeding researcher. "Just give them products that taste better, that they want to eat."

About the Author

Meghan Bartels is a science journalist and news reporter for Scientific American *who is based in New York City.*

Why People Oppose GMOs Even Though Science Says They Are Safe

By Stefaan Blancke

Genetically modified organisms (GMOs) have met with enormous public opposition over the past two decades. Many people believe that GMOs are bad for their health—even poisonous—and that they damage the environment. This is in spite of overwhelming scientific evidence that proves that GMOs are safe to eat, and that they bring environmental benefits by making agriculture more sustainable. Why is there such a discrepancy between what the science tells us about GMOs, and what people think? To be sure, some concerns, such as herbicide resistance in weeds and the involvement of multinationals, are not without basis, but they are not specific to GMOs. Hence, another question we need to answer is why these arguments become more salient in the context of GMOs.

I recently published a paper, with a group of Belgian biotechnologists and philosophers from Ghent University, arguing that negative representations of GMOs are widespread and compelling because they are intuitively appealing. By tapping into intuitions and emotions that mostly work under the radar of conscious awareness, but are constituent of any normally functioning human mind, such representations become easy to think. They capture our attention, they are easily processed and remembered and thus stand a greater chance of being transmitted and becoming popular, even if they are untrue. Thus, many people oppose GMOs, in part, because it just makes sense that they would pose a threat.

In the paper, we identify several intuitions that may affect people's perception of GMOs. Psychological essentialism, for instance, makes us think of DNA as an organism's "essence"—an unobservable and immutable core that causes the organism's behaviour and development and determines its identity. As such, when a gene is transferred between two distantly related species, people are likely to believe that this process will cause characteristics typical of the source organism to emerge in the recipient. For example, in an opinion survey in the United States, more than half of respondents said that a tomato modified with fish DNA would taste like fish (of course, it would not).

Essentialism clearly plays a role in public attitudes towards GMOs. People are typically more opposed to GM applications that involve the transfer of DNA between two different species ("transgenic") than within the same species ("cisgenic"). Anti-GMO organizations, such as NGOs, exploit these intuitions by publishing images of tomatoes with fish tails or by telling the public that companies modify corn with scorpion DNA to make crispier cereals.

Intuitions about purposes and intentions also have an impact on people's thinking about GMOs. They render us vulnerable to the idea that purely natural phenomena exist or happen for a purpose that is intended by some agent. These assumptions are part and parcel of religious beliefs, but in secular environments they lead people to regard nature as a beneficial process or entity that secures our wellbeing and that humans shouldn't meddle with. In the context of opposition to GMOs, genetic modification is deemed "unnatural" and biotechnologists are accused of "playing God." The popular term "Frankenfood" captures what is at stake: by going against the will of nature in an act of hubris, we are bound to bring enormous disaster upon ourselves.

Disgust also affects people's attitudes towards GMOs. The emotion probably evolved, at least in part, as a pathogen avoidance mechanism, preventing the body from consuming or touching harmful substances. We feel repelled by things that possibly contain or indicate the presence of pathogens such as bodily fluids, rotten meat, and maggots. This would explain why disgust operates on a hair trigger: it is better to forego an edible meal under the misguided assumption that it is contaminated, than to consume sickening, or even lethal, food that is erroneously thought to be safe. Hence, disgust can be elicited by completely innocuous food.

GMOs probably trigger disgust because people view genetic modification as a contamination. The effect is enforced when the introduced DNA comes from a species that is generally deemed disgusting, such as rats or cockroaches. However, DNA is DNA, whatever its source. The impact of disgust explains why people feel more averse towards GM food than other GM applications, such as GM medicine. Once disgust is elicited, the argument that GMOs cause cancer or sterility, or that they will contaminate the environment, becomes very convincing and is often used. Disgust also affects moral judgments, leading people to condemn everyone who is involved with the development and commercialization of GM products. Because people have no conscious access to the emotional source of their judgments, they consequently look for arguments to rationalize them.

Our cognitive analysis is not intended to debunk every anti-GMO claim a priori. A particular GM application may have unwanted effects, which can also be the case with a product of organic or conventional farming. The risks and benefits should be assessed on a case-to-case basis, regardless of the process. The current applications have been proven to be safe. One may take issue with the involvement of multinationals or be concerned about herbicide resistance, but these issues have to do with how GM technology is sometimes applied and certainly do not warrant resistance to the technology and to GMOs *in general*. The emotional and intuitive basis of anti-GMO sentiments however prevents people from making these distinctions.

The impact of intuitions and emotions on people's understanding of, and attitudes towards, GMOs has important implications for science education and communication. Because the mind is prone to distorting or rejecting scientific information in favour of more intuitive beliefs, simply transmitting the facts will not necessarily persuade people of the safety, or benefits, of GMOs, especially if people have been subjected to emotive, anti-GMO propaganda.

In the long run, education starting from a young age and specifically targeted at tackling common misconceptions might immunize the population against unsubstantiated anti-GMO messages. Other concerns can be addressed and discussed in the wider context of agricultural practices and the place of science and technology in society. However, for now, the best way to turn the tide and generate a more positive public response to GMOs is to play into people's intuitions as well. For instance, emphasizing the benefits of current and future GM applications–improved soil structures because herbicide resistant crops require less or no tilling, higher income for farmers in developing countries, reduced vitamin A deficiency, virus and drought resistance, to name a few–might constitute the most effective approach to changing people's minds. Given the benefits and promises of GM technology, such a change is much needed.

About the Author

Stefaan Blancke is a philosopher at Ghent University and co-editor of Creationism in Europe *(Johns Hopkins UP, 2014). His interests include pseudoscience and the way it reflects underlying intuitions. He has also published on the history of science, public understanding of science and science education, often from evolutionary and cognitive perspectives.*

Stop Arguing over GMO Crops

By Sarah Garland

The election of Joe Biden and Kamala Harris as president-elect and vice president-elect of the United States has been celebrated as a win for the scientific community. In his acceptance speech, Biden expressed that the American people have called on him and Harris as leaders, "to marshal the forces of science and the forces of hope in the great battles of our time." Climate change and the COVID-19 pandemic have featured heavily in this presidential campaign and are without question two of the great battles we are currently facing that require scientific expertise.

But this mandate to lead with science must extend to all challenges where science is needed, including agriculture. The Biden administration will undoubtedly listen to scientists when it comes to agricultural decision-making. The public that voted in the name of science must do the same.

Agriculture is inextricably linked to many global challenges we are now facing. Our current agriculture system is a major emitter of greenhouse gases (about a quarter of total emissions) while at the same time agriculture is dangerously vulnerable to harmful effects of the climate crisis. Extreme temperatures, floods, droughts and pest migrations are all examples of harsh environmental conditions exacerbated by climate change that are increasingly threatening crops worldwide. Amidst these pressures, the practice of deforestation to create new swaths of arable land must end in order to preserve biodiversity and maintain valuable carbon sinks.

In order for global agriculture to rapidly become more sustainable and climate resilient, we must have the power to responsibly use every tool at our disposal, including genetic engineering. No single solution is a silver bullet, and successful strategies to improve agricultural sustainability will likely involve a wide range of practices used together; experts have been clear about the need to combine low-tech and high-tech approaches for years. Current research in plant biotechnology focuses on developing plant traits related to climate resilience and sustainability such as drought tolerance, disease resistance and capturing carbon from the atmosphere. How and when different methods need to be used is for experts to decide, but the public must be open to biotechnology as a part of sustainable solutions.

For too long the rhetoric on genetic engineering in agriculture has been polarized, and reasoning has been clouded by emotion instead of based on evidence. There is fault on both sides for the existing antagonism. Immensely successful marketing strategies have been based on inciting fear and distributing false information. Large corporations have sewn mistrust. Promises were made that were never realized. But in 2020, the "GMO vs. organic" debate is outdated and unproductive.

Plant genetics research and development today has advanced far beyond the early GMOs of the 1990s. New technologies are redefining and revolutionizing genetic engineering in ways that couldn't have been imagined even only 10 years ago. Innovations will continue to emerge that do not fit old categories and old assumptions. Meanwhile, modern large-scale organic agriculture has far more in common with conventional industrial practices than the quaint farm pictured on the grocery store packaging suggests. Entrenched and misleading narratives on both GMO and organic production methods must be dismantled; the divisive rhetoric hinders progress that is urgently needed in order to create a more sustainable future.

The vast majority of the scientific community agrees on the safety and potential of genetically engineered crops. Over 150 Nobel Laureates signed an open letter supporting genetically engineered crops in 2016 that stated, "Opposition based on emotion and dogma contradicted by data must be stopped." The American public cannot have selective hearing when we say we will listen to scientists. We need to fight for the facts on genetic engineering and for the validity of the experts in this field the same way we fight for data and truth on climate change. We need to come together to focus on evidence to make decisions instead of letting old emotional battles divide us and block solutions.

President-elect Biden made a plea to a highly divided American people in his acceptance speech, "It's time to put away the harsh rhetoric. To lower the temperature. To see each other again. To listen to each other again. To make progress, we must stop treating our opponents as our enemy." The same message applies to the established factions in this nation regarding the use of genetic engineering in agriculture. The stakes are too high to let the antagonism continue any longer. We need to act in the manner in which we voted–supportive of science and hopeful for a more sustainable and unified future.

About the Author

Sarah Garland, Ph.D., is a postdoctoral research fellow at the Earth Institute, Columbia University.

Section 4: Saving Endangered Species

The American Chestnut's Genetic Rebirth

By William Powell

In 1876 Samuel B. Parsons received a shipment of chestnut seeds from Japan and decided to grow and sell the trees to orchards. Unbeknownst to him, his shipment likely harbored a stowaway that caused one of the greatest ecological disasters ever to befall eastern North America. The trees probably concealed spores of a pathogenic fungus, *Cryphonectria parasitica*, to which Asian chestnut trees—but not their American cousins—had evolved resistance. *C. parasitica* effectively strangles a susceptible tree to death by forming cankers—sunken areas of dead plant tissue—in its bark that encircle the trunk and cut off the flow of water and nutrients between the roots and leaves. Within 50 years this one fungus killed more than three billion American chestnut trees.

Before the early 1900s the American chestnut constituted about 25 percent of hardwood trees within its range in the eastern deciduous forests of the U.S. and a sliver of Canada—deciduous forests being those composed mostly of trees that shed their leaves in the autumn. Today only a handful of fully grown chestnuts remain, along with millions of root stumps. Now and then these "living stumps" manage to send up a few nubile shoots that may survive for 10 years or longer. But the trees rarely live long enough to produce seeds because the fungus almost always beats them back down again.

In its prime, the American chestnut was a keystone species, crucial to the health of a multitude of organisms in its ecosystem. Many different birds, insects and small mammals nested in its branches and burrowed into its bark. Bears, deer, turkeys, blue jays, squirrels and other animals ate the large, nutritious chestnuts. After losing so many mature chestnut trees, wildlife populations declined and became less diverse. The oaks that have since replaced the chestnut cannot support as many animals; the acorns they produce are only half as nutritious. And chestnuts once generated larger quantities of nuts than oaks do today, in part because they flowered after frosts that might have destroyed delicate buds.

The American chestnut also had great economic value. Its nuts can be used for food or ethanol fuel. Because the American chestnut grows quickly, has sturdy, straight-grained wood and is very rot-resistant, it provides excellent timber. In fact, if the chestnut were still abundant, most decks would likely be made from its wood instead of from pressure-treated lumber, which often contains heavy metals and other preservatives that endanger the environment and people's health when they find their way into soil and food. Last, the American chestnut has been an especially beloved tree, immortalized in poetry, songs, books, street signs, and the names of many schools, hotels and parks across the country.

We do not have to stand by as the American chestnut becomes a distant memory for most people. The culmination of decades of research suggests that science can restore the tree and all the resources it once offered people and wildlife. After a century of ineffective efforts to combat chestnut blight, two approaches are now meeting with some success. One strategy attempts to create blight-resistant American chestnuts with an ancient horticultural technique: hybridization. By mating American chestnuts with far smaller, fungus-resistant Chinese chestnuts, researchers "backcross" the resulting hybrids with other American chestnuts to Americanize the trees as much as possible while, it is hoped, keeping all the genes responsible for blight resistance. In addition to being rather imprecise, however, backcross breeding requires many generations and thousand of trees to produce individuals suitable for restoration.

For those reasons, my many collaborators and I are focusing on a second approach, which relies on altering the chestnut tree's DNA in a much more exact way than traditional breeding and which has the potential to produce more fungus-resistant trees more quickly. By borrowing genes from wheat and the Chinese chestnut, among other plants, and inserting them into the American chestnut's genome, we have created hundreds of transgenic trees, some of which defend themselves against *C. parasitica* as well as, if not better than, their Asian counterparts. If the U.S. Department of Agriculture, the Environmental Protection Agency, and the Food and Drug Administration approve our trees—which could happen as soon as five years from now—they will be the very first transgenic organisms used to restore a keystone species to its native environment.

Compared with other efforts to revive endangered or extinct species with genetic engineering and related biotechnologies—such as the proposed restoration of the passenger pigeon, thylacine and mammoth—the efforts to reinstate the American chestnut face far fewer hurdles and offer much clearer benefits. Unlike cloned mammoths and pigeons, trees do not require surrogate mothers, parenting or socialization. And as a massive organism that is home to many others, the American chestnut can improve the health of the forest more than any one animal.

Seeds of Salvation

Like many adults in the U.S. today, all I knew about chestnuts while I was growing up was what I learned from a certain iconic Christmas song. Yet in 1983, when I became a graduate student working with plant pathologist Neal Van Alfen, then at Utah State University, I began to develop a deep appreciation and sympathy for the magnificent chestnut tree and its demise at the hands—or rather the fungal fingers—of an exotic pathogen.

In 1989, when I had moved to the S.U.N.Y. College of Environmental Science and Forestry, Stan Wirsig of the American Chestnut Foundation approached my colleague Charles Maynard and me with a proposition. He wanted to complement the foundation's ongoing chestnut tree hybridization program with a new restoration project focused on genetic engineering, which was a cutting-edge technology at the time and promised a speedier and more precise way to create resistant American chestnuts. One of my tasks was to find a gene that could endow the trees with resistance to *C. parasitica* while Maynard and Scott Merkle of the University of Georgia developed the techniques that would allow us to introduce that gene to chestnut tree embryos–tiny bundles of swiftly multiplying cells that would eventually grow into adult trees. If everything worked as planned, the young trees would grow into sturdy adults with the ability to battle the fungus.

At that time, no one had ever tried to genetically engineer a tree to fight a virulent fungus, but we had a few clues about how to get started. Over the years researchers had learned some important details about how *C. parasitica* damages chestnut trees. The pathogen grows feathery lattices of fungal tissue called mycelial fans that produce oxalic acid, which eats through the tree's bark to make room for the fungal invasion. As the fungus wedges its way into the tree, a canker girdles the trunk.

Initially we focused on finding a way to weaken the mycelial fans. We knew that the immune systems of many plants and animals contain small chains of amino acids known as antimicrobial peptides (AMPs) that can disrupt fungal cells. Using AMP genes in the African clawed frog as a model, we assembled genes from scratch to produce AMP peptides that could fight *C. parasitica*. We hoped that if we could engineer the chestnut trees to produce even small amounts of these AMPs, they would make mycelial fans go slack and thereby render them benign. Such peptides are notoriously unstable molecules, though, so we needed a backup plan.

Around the same time, a then graduate student named Kim Cameron stopped by my office and dropped off a book summarizing many of the studies presented at the recent annual meeting of the American Society of Plant Biologists. When I read about a study conducted by Ousama Zaghmout and Randy Allen, both then at Texas Tech University, I had a eureka moment. The study described a wheat gene for an enzyme called oxalate oxidase (OxO), which breaks down oxalic acid–the very same caustic substance produced by the chestnut blight fungus. Even better, the researchers had worked out a way to introduce this gene into other plants. They put the gene into *Agrobacterium*, a microbe that can inject DNA into the command center of plant cells, and exposed plants to clones of that microbe. The resulting transgenic plants became resistant to an acid-spewing fungus known as *Sclerotinia sclerotorium*. Maybe we could do something similar with the American chestnut.

We could not test either approach on chestnuts at that point, because we were still figuring out how to grow the finicky chestnut in the laboratory. So we decided to achieve a proof of principle in a different tree–the hybrid poplar, which was well studied and often used in experiments. Haiying Liang, then a graduate student at the College of Environmental Science and Forestry, would deliver both the OXO gene and our AMP gene, and when the trees were old enough, we would infect them with *Septoria musiva*, a fungus that produces a good deal of oxalic acid and can cause leaf spot and canker diseases in hybrid poplars. Most of the trees treated in this way remained relatively healthy. We had made one tree fungus-resistant with genetic engineering. Now we needed to do it with the right tree and the right fungus.

While Liang was conducting the poplar experiments, Linda McGuigan, also then a graduate student at the college, set to work figuring out how to raise chestnut trees from embryos in the lab. Some plants, like carrots and petunias, are remarkably easy to grow in the lab. Provided with enough water, nutrients and certain hormones, they will grow new shoots and roots from a tiny piece of leaf, for example. The American chestnut was not one of these cooperative plants. McGuigan, building on the work of previous students, spent two and a half years learning how to successfully introduce the wheat gene into chestnut embryos using *Agrobacterium* and to subsequently shepherd the embryos into young adulthood in the lab. Usually the cluster of rapidly dividing cells that make up a chestnut tree embryo grow within the protective husk of a chestnut seed that has fallen to the ground, eventually pushing roots through the seed and into the soil and pushing green shoots toward the sun. McGuigan learned how to control lighting, humidity and temperature to mimic what would normally happen inside a chestnut seed and fine-tuned the delivery of various hormone cocktails at different stages of the miniature tree's early development to induce growth of roots and shoots.

In 2006 we were able to plant the first transgenic American chestnut trees in experimental fields sectioned off from the forest. It takes at least two to three years for the trees to reach a size at which we can challenge them with the blight fungus. We had attached the *OXO* gene to a promoter–a kind of genetic switch that controls how often a cell reads the instructions in a gene–to limit the production of OxO to certain tissues. We were hoping the resulting low levels of the enzyme would be sufficient to take on the fungus without causing any unwanted side effects. Unfortunately, we were mistaken. This first line of trees was not able to resist the fungus; they died a little slower than is typical but ultimately succumbed to their illness.

By 2012 we had designed a new promoter for the *OXO* gene and engineered a new line of trees that produced much more of the acid-degrading enzyme. Success! These trees evaded disease almost as well as the Chinese chestnut, which had evolved resistance on its own. We have now developed a way of gauging disease resistance by testing the leaves of chestnut trees that are only a few months old, so we no longer have to wait three years to see if our experiments are working. In this test, we make small cuts in leaves, infect them with fungus and wait for a circle of decaying tissue to spread from the wound. The smaller the spot of death, the more resistant the tree. Some of our newest trees, which make OxO in all their tissues and were planted in the field in 2013, appear to be even more resistant than the Chinese chestnut. We need to confirm this finding as the trees get older, but it appears that the gene we borrowed from wheat has exceeded our expectations.

People often ask us why we do not simply find the genes that make the Chinese chestnut resistant and use them instead of the wheat gene. When we first started our research, no one had thoroughly studied the Chinese chestnut genome, and it would have taken too much time and too many resources to locate the numerous different genes responsible for a complex trait like blight resistance. Each of those genes would contribute only a small portion of the tree's ability to battle the fungus, and any one of them would probably have been ineffective as a defense on its own.

At this point, however, scientists have identified 27 genes that might be involved in the Chinese chestnut's blight resistance–the fruits of a recent collaborative effort under the Forest Health Initiative between many researchers at the College of Environmental Science and Forestry, the University of Georgia, Clemson University, Pennsylvania State University, the U.S. Forest Service, North Carolina State University, the Connecticut Agricultural Experiment Station and the American Chestnut Foundation. So far two of these genes each appear to endow trees with an intermediate level of resistance. Testing is ongoing with the other candidate genes. Joseph Nairn of the University of Georgia has also given us copies of two other genes to test: one for a grape enzyme that helps to make resveratrol, which is toxic to fungus, and a pepper gene encoding an AMP that directly inhibits the growth of fungal cells.

Eventually we hope to fortify American chestnuts with many different genes that confer resistance in distinct ways. Then, even if the fungus evolves new weapons against one of the engineered defenses, the trees will not be helpless.

Going Out on a Limb

Today more than 1,000 transgenic chestnut trees are growing in field sites, mostly located in New York state. The next hurdle for American chestnut restoration involves the federal regulatory process. Before we can plant trees in the forest, the FDA, USDA and EPA will want to make sure that genetically engineered chestnut trees are not significantly different from typical trees in some unexpected way. As opposed to hybridized trees, which are genetically quite different from American chestnuts because they have large chunks of Chinese chestnut DNA, our transgenic trees have only a few new genes. Preliminary tests show that the roots of typical chestnut trees and engineered trees form the same kinds of symbiotic relations with helpful fungi and that similar communities of smaller plants grow underneath the canopies of both modified and unmodified trees. Likewise, the same insect species visit both transgenic and typical chestnut trees, and nuts from both types of trees have the same nutritional composition.

Once such tests are complete, we will petition the USDA, EPA and FDA for the same unregulated status that they give to genetically engineered crops. Here is where the American chestnut will introduce a new dilemma in the usual regulatory process. We are not growing a genetically modified organism on cropland for profit; rather we are producing trees for restoration without monetary gain. Like researchers working on golden rice enriched with a precursor of vitamin A, we are motivated by the public good–and the health of the forest. The EPA generally grants seed companies licenses to sell transgenic seeds, but in our case, we have no one to hold the license and nothing to sell. It is not clear what kind of alternative approval they would give us, but we are determined to set a precedent.

A final hurdle is public acceptance. Encouragingly, many people who are typically opposed to genetic modification make an exception for the American chestnut tree. Some people reason that because humans caused the demise of the chestnut in the first place, humans should fix it. Others are accepting because we are not seeking profit and are not patenting the trees.

Many people are also happy to learn that the environmental risks of American chestnut restoration are negligible. The chances of transgenic chestnut tree pollen spreading introduced genes to other plant species are very small. Pollen from one tree species can fertilize only the same species or a closely related one. The American chestnut has no closely related species in the northern part of its natural range. In the southern parts of its range, chinquapins occasionally cross with American chestnuts. But chinquapins are also infected by chestnut blight and would benefit from some genetic resistance. Ideally, some of the transgenic pollen will spread resistance to at least a fraction of the remaining American chestnut stumps that manage to flower, rescuing as much of their total genetic diversity as possible. If the stumps do benefit, they could spawn a blight-resistant population that, over the centuries, could return this once towering keystone species to its former glory in the eastern forests.

Chestnut blight is not the only enemy of biodiversity that genetic engineering can eradicate. We are losing the battle against many other exotic pests such as the hemlock woolly adelgid–a bug that sucks the sap from hemlock trees–and the emerald ash borer–a metallic green beetle whose larvae tunnel under the bark of ash trees–as well as the pathogens responsible for sudden oak death and walnut thousand cankers disease, to name a few. To turn the tables, we have to act quickly, and in most cases, traditional breeding techniques are just too slow to make a difference. Now, more than ever, we need genetic engineering in our toolbox to maintain diverse and healthy forests.

Completely restoring the American chestnut to its previous status as a king of the forest is a centuries-long endeavor. Once the chestnut trees pass regulatory and public approval, a good place to begin restoration is on reclamation lands. With the help of the Forest Health Initiative and Duke Energy, test plots are now being planted on mine reclamation sites. Other areas might include abandoned farmland and historic locations that once had abundant chestnut trees. And perhaps some individuals will want to have these iconic trees in their own yards. An old Chinese proverb says, "One generation plants a tree, the next generation enjoys its shade." In the case of the American chestnut, we are that first generation.

Referenced

Restoration of Threatened Species: A Noble Cause for Transgenic Trees. S. A. Merkle et al. in *Tree Genetics & Genomes*, Vol. 3, No. 2, pages 111–118; April 2007.

Transgenic American Elm Shows Reduced Dutch Elm Disease Symptoms and Normal Mycorrhizal Colonization. Andrew E. Newhouse et al. in *Plant Cell Reports*, Vol. 27, No. 7, pages 977–987; July 2007.

About the Author

William Powell is co-director of the American Chestnut Research and Restoration Program at the S.U.N.Y. College of Environmental Science and Forestry. He is a recent recipient of the Forest Biotechnologist of the Year Award from the Forest Biotechnology Partnership.

Could Genetic Engineering Save the Galápagos?

By Stephen S. Hall

On September 25, 1835, during the HMS *Beagle*'s sojourn to the Galápagos archipelago, Charles Darwin first set foot on what was then known as Charles Island. He found a colony of 200 to 300 inhabitants, nearly all political exiles sent there by Ecuador, aka the "Republic of the Equator," after a failed coup attempt. The lowlands did not much impress Darwin, with their "leafless thickets," but after trudging four miles inland and upward to a small, impoverished settlement in the highlands, he found "a green and thriving vegetation," cultivated with bananas and sweet potatoes, along with a group of islanders who, "although complaining of poverty, obtain, without much trouble, the means of subsistence." That was mainly because of the tens of thousands of giant tortoises that once prowled these islands. "In the woods," Darwin noted, almost as an afterthought, "there are many wild pigs and goats."

On the morning of August 25, 2017, Karl Campbell bounded off a twin-engine motorboat and onto the dock of that same humble island. Now known as Floreana, the island has 144 residents, half as many as in Darwin's time, and Campbell seemed to know them all. Dressed down in a baseball cap, blue jeans and gray T-shirt that read "Island Conservation," he ambled up to Claudio Cruz, at the wheel of a local bus (a converted truck with benches in the back), and exchanged some banter. He waved hello to Juanita and Joselito, who manned the Ecuadorian government's biosecurity checkpoint on the dock. He shouted out another "*Hola*" to the postmaster, popped his head into the community center to greet Myra and Holger, a farmer, and paused to catch up with Carmen, the woman who monitors the public bathrooms near the landing. His path up Floreana's one paved road was interrupted by salutations, chitchat, short jokes and the one-cheek kisses that are the custom in Ecuador.

Campbell, 42 years old at the time of the interview, is an Australian who had lived in the Galápagos Islands for 20 years; he's a gregarious and outgoing fellow, with a tendency to begin conversations with "All good, mate?" But the cheery demeanor and bonhomie he displayed that morning is an essential part of a massive scientific undertaking. Campbell has a Ph.D. in vertebrate pest management from the University of Queensland in Australia, and in 2006 he began working as an animal removal specialist for Island Conservation, an organization based in Santa Cruz, Calif., that is devoted to preserving biodiversity and preventing extinctions by removing invasive species from islands throughout the world. Campbell has been working on eradications in the Galápagos since 1997, including a 2006 campaign to remove all the feral goats and donkeys from Floreana. A decade later he's a project manager with Island Conservation, and the most ambitious project on its agenda is once again on Floreana: to eradicate every single rat and mouse on the island.

There are hundreds of thousands of islands in the world. "You can't work on all of them," Campbell says. Conservationists, according to Campbell, "are currently able to do 10 to 20 islands a year to rid them of mice. So which are the ones you should be working on most urgently? We basically draw up a list of places where we should be working to prevent extinctions." Topping that list, he says, is Floreana.

"Floreana has one of the highest endemicity rates in the Galápagos, the highest rate of extinctions due to the invasive species here and the highest rate–by far–of critically endangered species, which makes it one of the highest-priority targets not just in the Galápagos but in the world," Campbell says, in a spiel that has the polish and urgency of countless recitations to funders, journalists and probably every one of Floreana's residents.

Floreana is at the limit of feasible projects using current eradication tools. The island is large (17,253 hectares, or about 46,600 acres), and it is inhabited, which complicates the task enormously. It means having to explain the logistics and consequences of the entire project–not least of which is a plan to dump 400 tons of rodent poison all over the island. That is why, since 2012, Campbell and his colleagues, such as Carolina Torres and Gloria Salvador, have been visiting Floreana almost once a month, enduring the bumpy two-hour boat ride from the main island of Santa Cruz to meet with residents, describe their proposed project, and figure out the massively complicated steps needed to protect adults, children, livestock, water and endangered species from the effects of the poison.

Such eradications require almost military-scale logistics and precision, which is why Campbell has been desperately seeking an alternative to the blunt-force tools of current techniques. One of the most appealing, to his mind, is a controversial new form of genetic manipulation known as gene drive. Compared with the frustrations he endures every day on the Floreana project, he likens the technology to a magic wand out of *Harry Potter.*

The basic strategy of using gene drive in the conservation setting would be to tinker with the DNA of mice, using either the new gene-editing tool CRISPR or other tools of genetic manipulation, in such a way as to tilt the odds of sex inheritance; one example would be to produce offspring that would be exclusively male, eventually producing a daughterless population of mice. The elimination of females, of course, would create a reproductive dead end for that invasive species. Gene drive is far from a practical technology at this point, but Island Conservation has been working with molecular biologists in the U.S. and Australia to create these genetically modified mice, and Campbell has made no secret of his enthusiasm for the approach at recent scientific meetings. And that, in turn, may be why the National Academies of Sciences, Engineering, and Medicine, in a 2016 analysis on the potential benefits and risks of gene drive, included the example of daughterless mice among a series of potential scenarios where the technology might be applied. As the report noted, "Perspectives on the place of human beings in ecosystems and their larger relationship to nature–and their impact on and manipulation of ecosystems–have an important role in the emerging debate about gene drives." That debate, in a sense, has already begun on Floreana, where residents have been weighing the benefits and risks of a massive, albeit nongenetic manipulation of their precious ecosystem for the past five years.

Campbell is the first to acknowledge that the Galápagos will not be the first or best place to test gene drive in the field. But it may be the best place to think about the implications, good and bad, of gene drive in the context of species preservation. If, as a global community, we value the preservation and protection of biodiversity in the Galápagos (a value ratified by its selection as among the first World Heritage sites by the United Nations agency UNESCO), we also have to come to terms with the complexities and paradoxes of invasive-species eradication, which legitimizes the local elimination of certain animals for the benefit of other species–not least humans. As Campbell likes to point out, "No one comes to the Galápagos to see rats and goats and cats."

"Red in Tooth and Claw"

If the Galápagos Islands have become synonymous in the public imagination with ecological harmony and thrillingly pristine biodiversity, the reality is somewhat different. Yes, the massive tortoises are stunning, but where thousands of them once bulldozed the highlands of Floreana, there are now about two dozen–all imported from other islands because the local species went extinct. Yes, the fearless finches are charming and beautiful, but the Floreana mockingbird disappeared from the island around 1880, one of 13 species that have gone locally extinct. Yes, the sea turtles languorously swimming off La Lobería Beach are magnificent, but their eggs have been relentlessly poached by indifferent predators. All those iconic Galápagos species have been ruthlessly threatened by invasive species.

There is a darkness to the Galápagos paradise, and it has been there a long time, perhaps since Tomás de Berlanga, then the bishop of Panama, went off course and discovered the islands in 1535.

The first true invasive mammals on the islands were the pirates who frequented them in the 17th century, followed by sailors from whaling ships in the 18th century. These mariners brought in tow a malign ark of mammalian deplorables they introduced to islands that had been largely unperturbed for millions of years. If you want to be provocatively precise about it, the very first documented resident invasive species on Floreana was an Irish sailor named Patrick Watkins, marooned around 1805. He reportedly grew vegetables, which he bartered to visiting ships in exchange for rum (he was the model for a story by Herman Melville).

Three years before Darwin's arrival, a zoo's worth of invasive species had become entrenched on Floreana. It is no accident that in the scientific literature, the earliest date for many invasive species is 1832. That's when General José de Villamil, the first governor of the Galápagos Islands, arrived on Floreana to organize the penal colony. As Cruz–farmer, amateur historian, sometime bus driver and the largest landowner on Floreana–puts it, "He brought everything–goats, donkeys, cows, mules, horses, dogs, pigs, rats, everything." Similar animal importations occurred on other islands in the Galápagos during the 19th century, with devastating consequences on the local flora and fauna. Villamil brought the mules and donkeys to haul tortoises down from the highlands. At the time of his visit, Darwin reported that a previous ship visiting Floreana had loaded up on 200 tortoises in a single day (other ships reportedly collected as many as 700 apiece, according to Darwin).

Invasive mammals have wrought havoc on the ecosystem, in direct and indirect ways. Donkeys destroy tortoise eggs when they roll on the ground to cleanse themselves. Feral cats devour seabird chicks and snack on baby lava lizards, as do mice. Feral goats, in buzz cut fashion, chew through the native vegetation, removing the food that sustained the tortoise population for centuries and clearing the way for invasive plants such as guava, which has spread throughout the highlands. The Galápagos racer, once a common snake? Gone. More than 750 alien plant species and almost 500 alien insects have taken root in the Galápagos. As much as the islands have been a global classroom on evolution, they are also a reminder that nature is not static and that conservation sometimes alters nature to preserve it.

It has been the same story throughout the archipelago, though with some very odd chapters. In a 2012 compendium of "alien vertebrates" on the Galápagos, R. Brand Phillips, David A. Wiedenfeld and Howard L. Snell, all then affiliated with the Charles Darwin Research Station in Puerto Ayora on Santa Cruz, catalogued a rogue's gallery of 44 uninvited guest species, nearly half of them establishing feral populations. They ranged from obvious interlopers (goats, pigs, cattle, black rats) to an unwelcome menagerie of exotic animals. The Nile tilapia, a freshwater fish, turned up on the island of San Cristóbal in 2006; tree frogs have been spotted on two islands. Over the years nonnative visitors have included the mourning gecko, domestic ducks, cattle egrets, parakeets, peafowls and grackles. Three monkeys, of uncertain species, turned up on Floreana in the 1930s, and in 1937 one local entrepreneur tried to establish an ocelot colony on the island of Santiago. Ocelots!

Humans don't get a waiver from these waves of invasion, and their impact is increasing, too. In 1984 only 6,000 people total lived on five of the 129 islands and islets; more than 30,000 do today. And tourists? Three decades ago there were 20,000 a year; in 2016 there were 218,000. Just as more people began to come to the Galápagos to marvel at the local biodiversity, that biodiversity became increasingly threatened by the invasive species.

The Galápagos National Park Service, which controls 97 percent of the land in the archipelago, first attempted to eradicate goats on Pinta Island in 1971—an undermanned campaign that proved the adage in the eradication business that "a 99 percent success is a 100 percent failure." Only 10 goats remained on the island after the eradication program, recalls Victor Carrion, a former national park service official who participated in many eradication efforts. Within 10 years the number had climbed back up to 2,000. "The problem," Carrion says with a shrug, "was the final stage."

The Galápagos National Park Service began to develop more effective eradication plans in the late 1990s. Around this time, Campbell, then 22 years old and trying to decide what to do with his life, turned up in the archipelago. He had no particular affinity for the Galápagos—except, perhaps, that as a teenager back in Brisbane, he kept hundreds of pet birds in aviaries he built himself. In August 1997 he served as a volunteer on a goat-eradication project on the island of Isabela. Within a decade he would play a leading role in some of the most ambitious—and controversial—island eradication projects in the world.

The Devil We Know

Eradication is an ugly, euphemistic business. In 2004 the national park service and the Charles Darwin Foundation initiated a more systematic campaign to eradicate goats from the northern, uninhabited part of Isabela, the largest island in the archipelago. Two helicopters were used for aerial hunting; two or three hunters in each helicopter shot goats from the air, using semiautomatic 12-gauge shotguns and semiautomatic. 223-caliber AR15 rifles. After the first aerial sweep, ground hunters with specialized dogs went into heavily vegetated parts of the island to flush out goats that had survived the initial onslaught. In the final phase, beginning in March 2005, the eradication team deployed some 700 "Mata Hari goats" and "Judas goats."

Campbell's Ph.D. project was the development of the Mata Hari goat—a variation on the Judas goat, which was developed in the 1980s. Judas goats are outfitted with radiotelemetric collars. The animals are very gregarious, so hunters use goats wearing a wire, if you will, to find other goats. Mata Hari goats take the gambit one step further—they are female goats outfitted with hormonal implants that induce a permanent state of estrus, so that they seek and attract male goats. Mata Hari goats, needless to say, were not cooked up in the evolutionary hot pot of the Galápagos. Indeed, Campbell trained local hunters to perform field surgery on female goats—tying their fallopian tubes, terminating any pregnancies and inserting hormonal packs so that they were in constant heat, after which they were outfitted with radiotelemetry transmitters on collars so they could be traced. Once released, the Judas and Mata Hari goats tracked down the last holdouts. When all was said and done, Project Isabela killed 62,818 goats, at a cost of about $4.1 million. To hear Carrion tell it, the main complaint of the locals was that they didn't get any of the meat. "They said, 'We're hungry, and we need the food!'" he recalls. Even 100 percent success, in this case, wasn't enough—on at least nine occasions, according to Carrion, disgruntled locals deliberately reintroduced eradicated species, in part to protest local fishing regulations.

But the magnitude of the eradication campaigns in the Galápagos is staggering: 79,579 goats "removed" from Santiago, 41,683 from Pinta, 7,726 in San Cristóbal—in all, 201,285 goats have been "removed" from 13 islands (and you know it's a grisly business when euphemisms such as "removed" are used instead of "killed"). It's a pretty good bet that the tourists who flock to the Galápagos to swim with the sea turtles and follow the graceful arc of its storied birds are unaware that the islands have been turned into killing fields over the past two decades to preserve their famous biodiversity.

Even a modest rodent-eradication campaign illustrates just how tricky the traditional approaches can be. In 2012 the Galápagos National Park Service and collaborators began applying the rodent poison brodifacoum on the small, uninhabited island of Pinzón to eliminate rats, which had ravaged the eggs and hatchlings of giant tortoises for decades. The eradication was successful, and substantial numbers of tortoise hatchlings were reported on the island for the first time in a century. But the poison made its way into lava lizards, which in turn were eaten by endangered Galápagos hawks, resulting in at least 22 deaths because of brodifacoum poisoning (even though many of the hawks had been protected by "captive holding" for two weeks). In one instance, researchers found extremely high levels of rat poison in an owl carcass more than two years after the baiting.

And that brings us to the most ambitious island eradication in the Galápagos and perhaps anywhere in the world, an endeavor that everyone on Floreana refers to simply as the "*Proyecto*"–the Project.

The Devil is in the Details

There is one store in Floreana and one main road. As elsewhere in the Galápagos, the houses are simple cinder block constructions with corrugated metal roofs. If you go to one of the few restaurants in the island's single town, you better tell them ahead of time that you are coming: otherwise, they won't have enough food for you. The residents of Floreana are quiet-spoken, generous, subtly good-humored and deeply principled. Several years ago, when an entrepreneur from another island stiffed local workers out of their pay, no one on the island would serve him food, no one would rent him a room to sleep in and no one would speak to him. The entrepreneur's project collapsed. The island's quirky politics and fierce independence make such an endeavor socially daunting. As Campbell says, "It gets complicated real fast."

A recurring mantra in the recent National Academies report on gene drive–and, indeed, in almost every official white paper about genetic engineering in the past four decades–is the need for "public engagement." But that bloodless phrase does not begin to capture the passion and complexity of real projects in real circumstances. If eradications in general are hard, eradications on inhabited islands are *really* hard. That became clear to Campbell several years ago, during a small meeting with members of the Floreana community to discuss the *Proyecto*. One resident, adamantly opposed to the idea of having to remove livestock from the island, looked straight at Campbell and said, in unprintable language, "If you do this, I'm going to kill you." Campbell recalls the moment as "very conflictive."

The intensity of emotion does not seem entirely inappropriate, given the magnitude of the disruption. Since 2012 authorities in the Galápagos, with Island Conservation, have been formulating what they consider to be the most complex eradication plan of an inhabited island to date. It's not just the contentious adults on Floreana who make it complicated. It's children, pets and livestock, in addition to endangered birds and lava lizards.

Consider the staggering environmental risks of a "traditional," nongenetic eradication. To eliminate every rat and mouse from Floreana, the project calls for helicopters to drop some 360 million one-gram (0.035-ounce) pellets of brodifacoum—in Campbell's words, "Basically, systematically paint the whole island"—multiple times (drones are now being considered, after a pilot project on nearby North Seymour Island). To minimize potential health and environmental risks, the plan calls for extreme precautions. Water resources must be protected. Children may have to be removed from the island for up to six weeks. Pets will either need to be removed or restricted to domiciles or cages. Large agricultural livestock, such as cattle, pigs and horses, will have to be restricted in corrals (after the farmers of Floreana made clear that sending animals off the island for six months was not an acceptable option). Chickens will be housed in new covered coops, some of which have already been finished. Giant Galápagos tortoises in the Asilo de la Paz refuge will have to be temporarily restricted. Endangered birds will be trapped and held in specially built aviaries during the aerial baiting. In places off-limits to aerial baiting, such as buildings, homes or other structures, the eradication team will deploy traps and bait stations (the location of each bait station, in each home, has to be specified, and Carolina Torres, the lawyer for Island Conservation, is collecting signatures on draft agreements with each and every household). "A single pregnant female, or a single area missed, is a failure," Campbell says. "You need to get into every building, in every house, in every crawl space, in every closet, under every fridge to get every mouse."

The people from Island Conservation have taken the idea of "public engagement" to a new level. On a recent trip, Torres brought chocolates for Ericka Wittmer, a matriarch of one of the island's oldest families, and paid house calls to several island farmers to explain a legal issue involving contracts with the tenants who worked on parcels of their land. The organization recently provided paint for local homeowners to beautify their cinder block houses. When one resident expressed interest in starting a restaurant, Campbell and Torres encouraged her and promised to be customers. The organization has enlisted architects to design new chicken coops for the island's farmers; each unit will cost about $22,000. Campbell has learned the hard way that one-on-one relationship building is the best way to involve people in the decision-making process on such a delicate project. "If you do a town hall type of thing, they'll absolutely butcher you," he says. "Two or three people dominate the conversation, you don't know what other people think, and then afterwards, you have to spend a lot time dealing with the misinformation."

Despite initial reservations, Campbell says, most residents on Floreana support the eradication plan. In the highlands, Holger Vera, the farmer, stands amid a grove of orange trees, pineapple plants and other crops, lamenting the rapaciousness of the local rodents. They eat fledgling corn plants, he says; they devour pineapples; they eat the tubers of yucca. "Now they are even eating the sugarcane," he complains. "They are eating everything. But if we get rid of them, we can grow everything." Vera was initially skeptical about the *Proyecto*, Torres says, but he now sounds enthusiastic. Even if he has to board his seven dogs? "Yes, yes," he replies. Similarly, Cruz–who owns 80 cows, 130 pigs, more than 200 chickens, 10 horses and two dogs–agrees with the plan and the way it has been discussed with residents of Floreana. "We feel we are on the same page in terms of what's going on," he says.

"Essentially we have verbal agreements" from nearly all the residents, Campbell says. The plan still awaits final approval from Galápagos authorities. He believes the project could have been launched earlier if funding had been secured in a timely fashion. (Costs are expected to be $20 million overall, but funding hiccups have now delayed it until at least 2020; Campbell estimates that each year of delay costs $1 million.) Despite funding uncertainty, the reality of the *Proyecto* sunk in when seven orange, 20-foot shipping containers arrived in Floreana in 2017. They are intended to store uncontaminated livestock feed, or silage, for use during the rodent eradication; some farmers have already begun to store animal feed in the containers.

Pulling off a project this complicated is like managing a bureaucratic ecosystem—balancing the regulatory piece, the public engagement piece, the logistical piece, the funding piece, the poison mitigation piece. That's why Campbell thinks the Floreana project is "maxing out" the capability of traditional eradication tools. And that is why, not infrequently, he will say, "If we engaged the gene-drive technology, the conversations would be simpler, and the answers would be much more pragmatic."

The Devil We Don't Know

Campbell first became intrigued by the possibilities of gene drive in 2011, when he sat in on a conference call between biologists at North Carolina State University and officials of the U.S. Fish and Wildlife Service to discuss a possible genetic approach to control a runaway mouse problem on Southeast Farallon Island, about 20 miles west of the California coast, near San Francisco. John Godwin, a North Carolina State neurobiologist who studies animal behavior, had learned of the Farallon issue while skimming the Internet in 2011. He happened to be at a university with an established infrastructure dedicated to experimenting with—and considering the ethical implications of—genetic manipulation. Two of his colleagues, Fred Gould and David Threadgill, were already discussing the possibility of tinkering with the mouse genome in an attempt to create mice incapable of producing female offspring. Two other colleagues, Jennifer Kuzma and Jason Delborne, became deeply involved in how to engage the larger world of stakeholders—government regulatory agencies, animal management officials, bioethicists and, of course, the general public—in considering the prospect of releasing genetically altered animals into the wild. Kuzma and Gould serve as co-directors of the Genetic Engineering and Society Center at North Carolina State.

To make a long story short, Island Conservation joined forces in 2016 with other international groups to launch the GBIRd–Genetic Biocontrol of Invasive Rodents–program. GBIRd scientists are "cautiously investigating" genetic tools to preserve island ecosystems. The advent of the gene-editing tool CRISPR boosted efforts to develop an alternative approach to eradication. Those efforts gained traction in July 2017, when the Defense Advanced Research Projects Agency gave the North Carolina State group $3.2 million to pursue gene drives for mouse eradication on islands.

The basic idea of gene drive seems counterintuitive to anyone raised on the notion of Gregor Mendel's pea plants and the random inheritance of genes from parents. You usually have a 50–50 chance of inheriting a gene from one parent or the other. In rare instances, however, certain genes are favored, or "selfish"–they are inherited at much higher rates than random sorting would suggest. One such gene (technically, a region of the genome) exists in mice on chromosome 17; it is called the T-complex, and it is inherited at a rate of 95 percent. It might theoretically serve as a smuggler's bible, allowing a second gene to be quickly introduced in a population.

In an eradication scenario, researchers could theoretically attach a second piggybacking gene to the T-complex and essentially drive that second trait into the majority of offspring. One such mouse gene, known as SRY, determines male gender, so stitching it to a selfish gene would create more and more males (and fewer and fewer females) in each generation, until a mouse population would be daughterless. One of the basic requirements of gene drive is that the time between generations in the target animal is short; mice certainly qualify because their time between birth and reproductive maturity is 10 weeks. If the mice in the lab can be manipulated to pass along a desired gene, such as one to produce a single gender, and if those mice are reproductively successful in the wild, that gene could be rapidly driven into a population.

That's a lot of "ifs," but Threadgill, now at Texas A&M University, has been pursuing precisely that strategy in mice. This so-called daughterless breed could eliminate a native mouse population without environmental poison, without offshore animal relocations, without all the logistical nightmares entailed by the Floreana project. Paul Thomas, a biologist at the University of Adelaide in Australia, has been exploring the use of CRISPR to inactivate genes related to female fertility in mice, an approach that could be adopted to produce a population of entirely infertile females. In addition, Godwin, the neurobiologist, is testing whether an engineered mouse will pass sexual muster with wild mice (he is currently working with a batch transplanted from Southeast Farallon).

Species eradication is by no means the only application of gene drive. Target Malaria is an attempt to engineer mosquitoes so that they are incapable of transmitting malaria; the group, with funding from the Bill & Melinda Gates Foundation, has already begun community outreach efforts in Africa in anticipation of a field test. But Massachusetts Institute of Technology biologist Kevin M. Esvelt, who proposed a gene-drive experiment in mice on Nantucket Island, now feels a field trial is too risky. In the gene-drive game, the rule of thumb is that islands are the best place for a field test; smaller islands are better than larger ones, and uninhabited islands are better than inhabited ones. Campbell suspects the first field test of gene drive will involve mosquitoes and adds that the U.S., Australia or New Zealand would probably be the most appropriate venue because their regulatory infrastructures are sophisticated enough to assess new hot-button genetic technologies.

Eradications are controversial, genetic modification even more so. "There is no safe way to experiment with these technologies in the wild," says Dana Perls, senior food and technology campaigner at Friends of the Earth. Jane Goodall, Fritjof Capra and other conservationists called for a moratorium on the research in an open letter published in September 2016. Firing a shot across the bow of Island Conservation, the signatories said they were "alarmed that some conservation organizations have accepted funding for and are promoting the release of engineered gene-drive organisms into the wild."

The great fear is "unintended consequences"–that something unexpected and bad will happen. There is no question that gene drive, as the National Academies put it, "may have harmful effects for other species or ecosystems," and that alone warrants cautious and prudent development. But in previous public debates over genetic technologies, such as the battle over recombinant DNA in the 1970s, it was often difficult to separate legitimate concerns from exaggerated fears.

Back in the real world, during an excursion into the highlands of Floreana, Campbell and Torres led me to a freshwater spring–not far from the cave where the island's first settler, the drunkard Watkins, allegedly slept off his hangovers. As part of the project, the entire area surrounding the spring, which is already fenced off, will be covered with a tent, and special filters will be placed on the pipes to make sure no rodent bait gets into the system–even though brodifacoum is not water-soluble. Part of public engagement, Campbell said, is dealing with perceptions as well as legitimate fears. "You're working with people's perception of toxicants," he explained. "It's challenging to change people's perceptions of this, because they don't." One more reason, Campbell continued, that the genetic approach was more appealing. Then suddenly he changed the subject.

"Here we are," he said quickly, pointing to a rustle of vegetation inside the chain-link fence. "You see it? A rat!"

A pair of shiny, dark eyes briefly appeared amid the leaves. Campbell identified it as *Rattus rattus*–the black rat, which is known to eat the eggs and hatchlings of Galápagos petrels and giant tortoises. Like rats everywhere, it disappeared quickly–a sentinel of an inevitably larger population and a larger covert threat to what Campbell calls "species on the brink."

"The Stranger's Craft or Power"

Every stroll in the Galápagos is a nature walk, and each living creature tells a conservation story–some with happy endings, some not. During our last day on Floreana, a number of these stories began when Campbell's keen eye alighted on the animals that make this landscape so beloved–and beleaguered.

During breakfast, a cactus finch stalked our table. Its strong black and yellow beak had evolved to be larger and stronger, Campbell explained, to crack the unusually large and hard seeds of the local Opuntia cactus on Floreana; the cactus, in turn, is evolving even larger and tougher seeds to thwart this poaching–a reminder that evolution is not a textbook concept but an ongoing process. Moments later Campbell spotted a mouse darting behind a hunk of lava. As we finished our meal, another invasive species made an appearance–the sleek, black, smooth-billed ani (pronounced "Annie"). An example of old-school unintended consequences, farmers introduced the bird to the Galápagos in the 1960s in the belief that it could control ticks that afflicted cattle; it did not live up to its billing, so to speak, but it has exploded in numbers as an invasive species.

Later, on a walk to La Lobería Beach, Campbell pointed out fresh tracks of feral cats in the sand; they devour juvenile marine iguanas and lava lizards. ("The small ones have zero chance of getting away," he said.) Near the head of the beach, he indicated the gnawed-off limb of one of the *Opuntia* cacti. When rodents chew down the cacti, he explained, the plants fail to flower or bear fruit–eliminating a crucial source of sustenance for tortoises and mockingbirds, especially in the dry season, and depriving finches of nesting sites. And we paused to admire several magnificent sea turtles temporarily trapped in a lagoon during low tide. Their eggs and hatchlings, too, provide tasty meals for rats and cats.

It was Darwin's 20th-century bulldog, Richard Dawkins, who revived poet Alfred, Lord Tennyson's phrase "Nature, red in tooth and claw" to describe the noir side of natural selection–nature's game is not always pretty, and the postcard-perfect ecology of a place like the Galápagos often conceals a darker, more unsentimental interaction of predator and prey–an interaction whose delicate balance humans have repeatedly perturbed, whether by introducing invasive species or by attempting to atone for those ill-conceived introductions with literally toxic remedies. And now, on the horizon, we may have to decide whether to use futuristic techniques of genetic modification to restore the islands to an earlier, more pristine state.

For what it is worth, a small sampling of opinion on Floreana did not betray much local concern about the potential applications of gene drive, although it is not clear how well understood these technologies (and their potential risks) are. Vera shrugged off any worries and said he would have no problem with a genetic solution to the rodent problem. Ingrid Wittmer, another descendant of one of the earliest families on Floreana, shook her head no when asked, instead expressing concern about the fate of the short-eared owl once its main food source, mice, was eliminated during the *Proyecto.* Cruz, whose father emigrated to the island in 1939, when the population numbered 11, offered a farmer's perspective to the idea of daughterless mice: "It's like artificial insemination in cattle," he said. "If you want females, you use the semen for females. It's the same thing."

"For me, these are issues we've created, and to sit back and do nothing, there's going to be grave consequences," Campbell said. "We know where things are heading. To actually *not* do something is... is just irresponsible. If you have the tool, and you don't use it, *you're* culpable."

We don't have the tool yet. But if the craft of molecular biology eventually captures the power of gene drive, and it is used to manage invasive species in the Galápagos or any island, it is worth remembering that almost every ecological catastrophe visited on the planet's living laboratory of evolution has come at the hands of humans. The goats, the donkeys, the rats, the cats, the pigs, the mules, the mice and, yes, even those short-lived ocelots arrived with human help, on human boats, through human agency.

In a wry observation that resonates nearly two centuries later, Darwin remarked in his journal that while birds in England had developed a well-earned distrust of humans, the birds in the Galápagos "have not learned [such] a salutary dread." He went on to offer what might serve as cautionary words about 21st-century science and gene drives in particular. "We may infer from these facts," Darwin wrote, referring to the lack of fear in birds, "what havoc the introduction of any new beast of prey must cause in a country, before the instincts of the indigenous inhabitants have become adapted to the stranger's craft or power."

About the Author

Stephen S. Hall is an award-winning science writer and regular contributor. He is author, most recently, of Wisdom: From Philosophy to Neuroscience *(Knopf, 2010).*

De-extinction Company Aims to Resurrect the Tasmanian Tiger

By Kate Evans

The thylacine has long been an icon of human-caused extinction. In the 1800s and early 1900s, European colonizers in Tasmania wrongly blamed the dog-sized, tiger-striped, carnivorous marsupial for killing their sheep and chickens. The settlers slaughtered thylacines by the thousands, exchanging the animals' skins for a government bounty. The last known thylacine spent its days pacing a zoo cage in Hobart, Tasmania, and died of neglect in 1936.

Now the wolflike creature—also known as the Tasmanian tiger—is poised to become an emblem of de-extinction, an initiative that seeks to create new versions of lost species. Colossal Biosciences, a Texas-based de-extinction company that made headlines last September when it revealed that it planned to bring back the woolly mammoth, announced today that its second project will be resurrecting the thylacine.

Australian scientists have been hoping since 1999 to use emerging genetic technologies to try to bring the thylacine back from the dead. When the species went extinct, Tasmania lost its top predator. In theory, reintroducing proxy thylacines could help restore balance to Tasmania's remaining forests by picking off sick or weak animals and controlling overabundant herbivores such as wallabies and kangaroos, some researchers say. But early attempts at cloning the animal from museum specimens' DNA failed, and the effort has not attracted significant funding—until this year.

Colossal Biosciences, co-founded by Harvard University geneticist George Church and tech entrepreneur Ben Lamm, is working with the University of Melbourne's Andrew Pask, who has already sequenced most of the thylacine genome. The thylacine is the perfect candidate for de-extinction, Pask says, because it died out relatively recently, good-quality DNA is available, and its prey and parts of its natural habitat still exist.

In March his team established the Thylacine Integrated Genetic Restoration Research Lab with a philanthropic gift of five million Australian dollars (about $3.6 million). Colossal is providing "more than that" sum, Pask says–he won't divulge exactly how much–as well as access to equipment, another dedicated thylacine lab in Texas and a large team of researchers.

With this partnership established, Pask now says it's reasonable to expect to have "a de-extincted thylacine-ish thing" in a decade. That first iteration might be "90 percent thylacine," he says, though the ultimate goal is more like 99.9 percent. Eventually–after many years of monitoring the engineered animals in a large enclosed area–Colossal's goal is to release a viable, genetically-diverse population of perhaps 100 proxy thylacines into the wild.

To resurrect the woolly mammoth, Colossal's researchers plan to introduce mammoth genes into the genome of the Asian elephant, its closest living relative. They will then try to create an embryo carrying that modified DNA that could gestate in an African elephant "surrogate" or an artificial uterus. The resulting creature would not be a mammoth per se but rather a cold-adapted "Artic elephant" with small ears, shaggy hair, a domed forehead and curved tusks, Lamm says. Yet if he showed the creature to his grandmother, she'd say "that's a woolly mammoth," he adds.

Already, Lamm says, Colossal has fully sequenced the Asian and African elephant genomes, collected more than fifty mammoth genomes, and begun making edits to elephant cells–but he thinks the thylacine could turn out to be easier to revive than the mammoth because of the gestation times involved. Both projects still face many hurdles, however.

For the thylacine, the first task is to complete the sequencing of the animal's genome. Pask's lab has about 96 percent of it down, but the final 4 percent is the trickiest, he says. "It's like doing one of those horrible puzzles that's all baked beans or all blue sky. Every bit looks the same, and we're trying to figure out how it goes together."

Next the researchers will compare the genome of the thylacine to that of one of its closest living relatives: the fat-tailed dunnart, a mouse-sized marsupial that is relatively abundant and copes well in captivity. Using CRISPR gene-editing technology, the scientists will engineer the dunnart's genome to more closely resemble the thylacine's.

The researchers have already figured out how to re-program dunnart skin cells into stem calls, and are currently testing them to see whether they're capable of generating an entire embryo—something that hasn't yet been done in marsupials, which develop differently from placental mammals such as humans and mice. Once they've fine-tuned the recipe, they'll be able to use the stem cells to create a gene-edited living embryo they can insert into either a dunnart mother or an artificial marsupial womb, which they would have to invent.

Thylacine pregnancies are estimated to last just a few weeks, compared with 22 months for mammoths. And like other newborn marsupials, the baby thylacines would be little larger than a grain of rice, so even a diminutive dunnart mother could nourish them in her pouch at first. But Lamm says Colossal will work on developing a synthetic pouch, as well as a marsupial milk formula appropriate for each stage of development.

Collectively, these new marsupial reproductive technologies could become crucial tools for the conservation of extant species such as koalas or numbats, Pask says. "There is absolutely no way I would have the millions that I have now for marsupial conservation if I [wasn't] trying to bring the Tasmanian tiger back," he says.

Other scientists are considerably less optimistic about the project. Mammal expert Kris Helgen of the Australian Museum, who worked on sequencing the thylacine's mitochondrial genome in 2009, thinks altering the dunnart's DNA to truly resemble a thylacine's will be an impossible feat. The two species are separated by as much as 40 million years of evolution, he says. Thylacines are so unlike other animals that they're in their own taxonomic family, just as dogs are in one family of mammals and cats—from tigers to tabbies–are in another. Turning a dunnart into a thylacine, Helgen says, would be the equivalent of editing a dog's genome until the resulting animal looked like a cat. (Mammoths and elephants are far more closely related.)

Even if Colossal could overcome the technical challenges involved, the prospect of resurrecting the thylacine raises ethical concerns, according to Carol Freeman, an animal studies researcher at the University of Tasmania. "The whole discourse is about bringing this animal back, but the welfare of the individual animals isn't really talked about," she says. Both dunnarts and almost-thylacines would inevitably suffer in the course of these experiments, which "cannot be justified for such an uncertain result. It would be many years, if ever, that cloned thylacines could have anything like the life they may have had–and deserve–in the wild."

If the scientists do get to the point where they have actual living thylacines in hand, Pask says they would consult the public, including Indigenous communities, about any release. But Bradley Moggridge, a Kamilaroi environmental scientist at the University of Canberra in Australia, says Indigenous Australians should be involved now–especially Tasmanian Aboriginal peoples, who were themselves hunted by white settlers in the 19th century. "They may have ideas; they might need to get [their traditional lands] ready for this species. That could take a long time," he says. Discussions between the Colossal team and Indigenous Australians could be beneficial for everyone, Moggridge says. Aboriginal ecological knowledge about the thylacine would have been encoded in stories and songs, and de-extinction could reignite some of them, but the researchers "need to start those conversations now."

Other critics worry the glamour of de-extinction will rob attention and funding from conservation projects. One study in 2017 found that allocating sums to existing endangered species programs rather than giving the same amount of money to de-extinction efforts would see about two to eight times as many species saved. "It's better to spend the money on the living than the dead," lead author Joseph Bennett of Carleton University in Ontario told *Science*.

The idea that science could restore the thylacine "is just so lovely; it captures the imagination," says Helgen, who once made a pilgrimage to visit every museum specimen of the animal in existence. "But the thylacine is extinct in Australia and in Tasmania, and there's no way to bring it back." Some species are simply gone forever because of how unique they were, and the thylacine is one of them, he says. "A few million dollars [are] not going to give us an escape hatch from extinction."

About the Author

Kate Evans is a science and nature writer based in Raglan, New Zealand. She is a regular contributor to New Zealand Geographic, *and her work has also appeared in the* Guardian, BBC, National Geographic, *the* Washington Post, bioGraphic *and* Undark.

To Restore Biodiversity, Embrace Biotech's 'Intended Consequences'

By Paul Robbins, Ryan Phelan

In December of 2020, when scientists managed the incredible feat of cloning the endangered black-footed ferret, they took a leap toward the renewed global priority to combat climate change and biodiversity loss. The cloning success both fulfilled the founding vision of Earth Day and frightened its strongest proponents. By using biotechnology to achieve one of conservation's most critical goals–to restore genetic diversity to a species faced with a limited gene pool–conservationists made a step forward in saving a beloved species. But they did so by "tinkering" with the core mechanisms of life itself. This effort realized the vision of Earth Day founder and Nelson Institute namesake Gaylord Nelson, to foster "an environment of decency, quality and mutual respect for all other human beings and all other living creatures." Would Nelson have recognized and valued this advance? We believe so. To restore biodiversity, we will have to innovate.

While none of this obviates the need for traditional conservation measures, like habitat protection and land management, we can and we should address the looming extinction crisis with the same bold and rapid approach used to fight COVID-19. A creative ethos and an open mind can release the power of new technologies.

Together, we have years of conservation experience. One of us is an environmental researcher and the other the co-founder and executive director of Revive & Restore. We understand the importance of supporting and restoring biodiversity, and we are teaming up with the world's leading molecular biologists, technologists, conservation biologists, conservation organizations, ethicists and thought leaders to call for "Intended Consequences" to help us safely use all of the available tools that can provide the advantage we need to turn the tide on species loss.

Intended Consequences is a new, inclusive, ethical and rational framework that will help us envision bold conservation interventions and safely leverage biotechnology to win the race against extinction.

Some individuals worry about the *unintended* consequences of intervening with nature, including the use of genetic technology as well as traditional conservation restoration. However, alarming biodiversity loss tells us we must be more focused on the game-changing positive impacts that will result from a focus on Intended Consequences. If we fret endlessly about unintended consequences and wallow in uncertainty, we will inevitably witness the chilling result as a mass extinction plays out. Our current pivot point asks of us boldness and action as we consider biotechnology solutions and weigh the consequences of doing nothing.

The American chestnut, for example, will not survive without intervention. Prior to the industrial revolution, these trees formed endless stands in the Eastern forests of North America. By the 1940s a nonnative fungal blight killed an estimated four billion trees nationwide. As Eastern forests lost the American chestnuts, smaller trees established denser stands. This shift resulted in a new ecological state characterized by impoverished habitats, shrunken wildlife populations, inferior forest products and reduced biodiversity.

Towards the end of the 20th century, a team of innovative scientists began experimenting in the lab. They added a gene from wheat to the otherwise unaltered American chestnut genome. The resulting transgenic tree is no freak of nature. It is a 100 percent American chestnut that now produces an enzyme that degrades the blight's toxin. Because of this single additional gene, it can coexist on landscapes where the invasive fungus also thrives.

This seemingly radical solution is an elegant example of the potential for biotechnology to enable species-saving interventions. American chestnut fans want seedlings to plant in their yards, and the Eastern Band of Cherokee Indians in North Carolinahas signed onto plant blight-resistant chestnut trees on tribal land. But this early enthusiasm is only part of the process.

Responsible intervention must be guided by the Intended Consequences framework, including a thorough risk assessment and detailed studies to test how the proposed intervention could affect the ecosystem. We now know that when transgenic chestnut leaves fall into forest pools, they are safe for wood frog tadpoles to eat. The Federal regulatory system is reviewing the American chestnut project now; if it passes muster, it will be the first to demonstrate how carefully considered genetic interventions can enable coexistence in the wild.

With both the American chestnut and the black-footed ferret, 21st-century conservation solutions began in the lab. Responsible genetic interventions resulted from pairing new biotechnologies with decades of natural history knowledge and careful research. Despite the wariness cultivated by fictional horror stories, biotechnology is simply one more tool in this earnest race against extinction. The American chestnut is a critical proof of concept, and the black-footed ferret project is underway now, but we need these positive outcomes for all endangered species. We want to see resilient wild populations flourishing in nature.

Perhaps, in the not-too-distant future, we will see biotechnology applied to help save coral reefs. Scientists are already exploring how genetic interventions could be used to adjust coral temperature tolerance. And with Intended Consequences in mind, Revive & Restore is designing the Advanced Coral Toolkit to develop new tools, including stem cells, probiotics and rapid diagnostics, that will increase our options for reef restoration. If we do not intervene, we could lose coral reefs forever.

Conservation is finally ready to embrace the innovative spirit that drives problem-solving in other fields. An interdisciplinary group of scientists and conservationists from around the world agreed, and together they recently authored the Intended Consequences Statement to provide an initial framework for responsible conservation intervention that follows in the footsteps of the black-footed ferret and the American chestnut tree.

The Intended Consequences framework incorporates the lessons learned from decades of successful conservation work and redirects us away from despair and toward an optimistic future, encouraging us to imagine solutions to seemingly intractable conservation problems and inspiring us to act. It's something we can all agree on.

This is an opinion and analysis article.

About the Authors

Paul Robbins is dean of the Nelson Institute for Environmental Studies at the University of Wisconsin–Madison.

Ryan Phelan is co-founder and executive director, Revive & Restore.

Section 5: Questions and Controversies

The First Tinkering with Human Heredity May Happen in the Infertility Clinic

By Stephen S. Hall

Kyle Orwig has been itching to do an experiment that would, in his words, "piss people off." Orwig, a professor at the University of Pittsburgh, is an expert on the intricate biology of sperm cells—in particular, how specialized "stem" cells located in the male testes produce sperm. Every so often, however, a genetic flaw prevents these stem cells from completing this process, thus rendering the male infertile. The experiment Orwig has in mind is to use gene-editing technology to fix this flaw in the sperm-forming stem cells and then transplant them back into infertile mice, thereby demonstrating a potential treatment for male infertility.

It sounds simple enough, and, according to Orwig, it would be relatively straightforward to try—indeed, he has been transplanting sperm-forming stem cells into mice for 20 years. The consequences, however, could be momentous. The kind of experiment Orwig is contemplating would, if successful, push society right up to the brightest red line in contemporary biology: altering the genetic text of the human species in a way that is passed down to future generations.

If shown to be safe, effective and ethically acceptable, germ-line modification would confer unprecedented power on scientists—the power to edit the susceptibility to disease out of our species' DNA, for example, but also the power to manipulate human inheritance and "improve" the species, an aim that darkly harkens back to the eugenics movement of the early 20th century that reached its nadir in Nazi Germany.

Orwig, a broad-shouldered, burr-headed Oregonian who seems genial and determined, has no intention of crossing ethical lines. But he is something of a provocateur. By demonstrating that infertility in mice could be cured with a modest amount of genetic tinkering, he hopes to trigger a wider awareness that editing human genes is not an abstract, long-term technical challenge, as some have suggested, but a near-term possibility with practical medical consequences. Which is why Orwig recently told a colleague, "Let's just do this and piss some people off. Show them that it's possible, so nobody can say it's impossible. And get people talking about it."

The issue of germ-line modification has assumed great urgency in the past two years because of a powerful gene-editing tool called CRISPR/Cas9, which allows scientists to alter the DNA of any organism–including, potentially, humans–with unprecedented precision and ease. In April 2015 Chinese researchers reported the first attempt to edit the genes of human embryos. The headlines–"Embryo Editing Sparks Epic Debate," in *Nature*; "Eugenics Lurk in the Shadow of CRISPR," in *Science*–signaled broad social unease. In the alarmist shorthand of media accounts, the possibility of gene editing triggered fears of "designer babies" and "genetic enhancement."

The humble sperm cell, however, is a less controversial target. And whereas editing genes in an embryo remains a big challenge, many experts believe that approaching germ-line modification upstream of the embryo, in the sex cells that merge to form a zygote, is easier and arguably safer. Once you modify these cells, however, you are essentially modifying the human genome because the changes are permanently inscribed in the genetic text of the embryos they create. Orwig is among a handful of biologists who have experience genetically modifying and transplanting spermatogonial stem cells, the testicular cells that churn out sperm.

The field of reproductive medicine has a well-established track record for pushing headlong into the clinic with technological innovations. Infertility is also big business. If Orwig were to demonstrate in animals that a simple genetic fix is possible, it would be a tempting procedure for the tens of thousands of men who cannot make their own sperm, for whom options are currently limited, as well as for the in vitro fertilization (IVF) industry, which did an estimated $2 billion in business last year in the U.S. (and perhaps 10 times that amount worldwide).

Before administering any treatment, there must be proof that it works and causes no undue harm. Scientists would require such proof before even contemplating creating a human being with edited genes. Animal versions of those experiments are already under way, however, and the red line may be crossed soon. It could happen in China, where researchers have already taken the first tentative steps in editing human (albeit nonviable) embryos. It could happen in the U.K., where the government has legalized a form of germ-line modification known as mitochondrial replacement therapy and, this past February, approved gene-editing experiments on human embryos. And it could happen in an IVF clinic anywhere, building on recipes developed in laboratories such as Orwig's.

"This isn't theoretical," Orwig says. "The mouse is here, and the human is not too far in the future. The pieces are already in place."

Keep Calm and Carry On

The current debate over germ-line modification may feel familiar, but it treads on fundamentally new ground. Scientists began to acquire the Promethean ability to rewrite the language of heredity in the early 1970s, when biologists discovered that they could crudely cut and paste DNA with the use of enzymes harvested from bacteria, a technique called recombinant DNA. That advance caused unease about dangerous, genetically engineered microbes escaping the lab, prompting a voluntary—and unprecedented—moratorium on recombinant DNA research in 1974 and a historic meeting in 1975 of scientists at the Asilomar conference in California. Prominent molecular biologists such as David Baltimore, then at the Massachusetts Institute of Technology, debated the safety of the new technology, which led to federal guidelines governing the research. The Asilomar meeting was correctly viewed as a watershed cultural moment: Michael Rogers published a richly detailed account of "The Pandora's Box Congress" in *Rolling Stone*, and by the time the guidelines were in place, biotechnology emerged as one of the transformative industries of the 20th century.

Although society applauded the scientific community's decision in 1974 to hit the pause button on the headlong rush of research, many scientists felt it was an overreaction to hypothetical safety concerns. James D. Watson, co-discoverer of the DNA double-helix structure, called it "senseless hysteria."

Since Asilomar, controversial biology has often landed in the public square with a bang, prompting big meetings with messy background noise. When the National Academy of Sciences debated recombinant DNA in 1977, protesters opposed to genetic engineering unfurled a banner quoting Adolf Hitler: "We Will Create the Perfect Race." A meeting on human cloning in 2001 became a media circus. Maverick IVF doctors vowed to clone human babies. Television crews followed the would-be cloners everywhere (including to the bathroom). *Wired* magazine, in 2001, proclaimed on its cover: "Someone Will Clone a Human in the Next 12 Months."

This time around there is palpable unease among scientists, but they are also wary that another self-imposed moratorium might hold up progress. The result? Another meeting. Last December the National Academy of Sciences and the National Academy of Medicine co-hosted an international "summit" in Washington, D.C. (with the Royal Society and the Chinese Academy of Sciences). Baltimore acknowledged that altering human inheritance had remained "unthinkable" because of the cumbersome and inefficient first-generation tools of genetic engineering. "Over the years, however, the unthinkable has become conceivable," he said, "and today we sense that we are close to being able to alter human heredity." The overriding question, Baltimore continued, is, "How, if at all, do we as a society want to use this capability?"

The answer, to anyone who sat through the three-day meeting, as I did, seemed to be: we're not sure, but there's plenty of time to think it through. Numerous talks, including a plenary address by genome scientist Eric Lander of the Broad Institute of M.I.T. and Harvard University, stressed the technical hurdles and the lack of compelling medical needs for human germ-line modification in the near future. "It might, might, might be a good idea that before we make permanent changes to the human gene pool," Lander warned, "we should exercise considerable caution."

The organizers deftly steered clear of an Asilomar-like moratorium. Baltimore read a carefully worded statement from the meeting organizers acknowledging that it would be "irresponsible" to pursue human germ-line editing in the clinic at this time. At the close of the summit, he went on to explain that the organizers had deliberately avoided calling for a ban or moratorium. "Neither word did we want to use," he said. "Neither word have we used." Basic research could, and should, proceed unimpeded, but the public need not be concerned about imminent developments: human applications of germ-line editing were implausible, unnecessary, unwise and certainly not right around the corner.

That is not the way everyone in the scientific community sees it. The organizers of the Washington meeting framed the issue as "when, if at all." But a different word often crops up in private conversations with biologists when you ask about the prospects for germ-line editing. That word is "inevitable."

The Timeline

Some scientists viewed the National Academies meeting as an effort to "reinforce the status quo," according to biologist George Church of Harvard Medical School. "They basically want the public to calm down," he says. "That was their goal. And no matter what we said, that was going to be the goal. I don't want to stir [the public] up, and I don't want to calm them down, either. I want them to have an accurate view of where things are going." And the public needs to start thinking about gene-editing the human genome now, Church says, because science is already bumping up against the red line.

Despite a thicket of international provisions regulating human embryo research, Church and others believe that the creation of gene-edited sex cells in the test tube (the technical term is "in vitro gametogenesis," or IVG) has made great strides in recent years without attracting the same public scrutiny, or provoking the same ethical discomfort, as gene editing in embryos.

"In terms of the technology, that's ready to go now," says I. Glenn Cohen, a bioethicist at Harvard Law School. "In vitro gametogenesis is much closer to the mark than any other way to do it." And Ina Dobrinski, an expert in reproductive biology at the University of Calgary who works on gene editing of large animals such as pigs, adds, "Theoretically, we can do it. Practically [speaking], nobody is even touching it because of the ethical issues."

If human germ-line editing is inevitable, despite the ethical concerns (and indeed legal prohibitions in many countries), how might it happen? Speculation among biologists has become something of a parlor game, but I turned to Church, a card-carrying futurist, to lay out a plausible scenario. He was happy to oblige.

Church sees the germ-line Rubicon being crossed because sperm do not seem to arouse the same ethical passions as embryos or even egg cells. (Bioethicist Cohen agrees: "People do not believe that masturbation is genocide.") He also thinks gene therapy, not CRISPR per se, will set the table for this momentous change because it is already accepted: the U.S. Food and Drug Administration has allowed many trials of gene therapy in somatic (nongerm-line) cells. "Gene therapy is happening in young children already and will happen in younger and younger children," Church says.

In a highly publicized case last fall, for example, British researchers used gene-editing techniques to alter the immune cells of an infant battling leukemia. And the leap to germ-line gene therapy, according to Church, will occur not in human embryos but in the most proletarian, plentiful and expendable cell in the human body: sperm. Gene-editing sperm will spare couples the agony of destroying IVF embryos that, on preimplantation screening, are shown to possess variants that forecast some devastating single-gene disorders, he believes. "Maybe half the people in the U.S. already feel that they're not comfortable killing embryos, but I think people would be comfortable genetically altering sperm," he observes. "No embryos die."

Two obvious targets, Church adds, would be single-gene disorders (such as Tay-Sachs disease) and infertility. "You could also do it in human spermatogonial stem cells," he says, referring to the specialized adult stem cells in male testes that generate sperm cells—millions on millions of mindless, headlong swimmers—every day. "People don't really care about spermatogonia. Most people don't even know how to pronounce it. So they're going to let you mess around with them, right?" Church continues. "You will be able to do all kinds of things to show that they are functionally normal—that you've taken the sperm that can't swim, and that you've now taken their stem cells and made sperm that can swim. You can test that in a lab without any eggs being involved. And then in the fertility clinic, the dad will say, 'Hey, those are pretty awesome sperm. Let's try them out and see what they can really do.' And I don't know who's going to stop them from doing that."

As for the timeline, Church says, "I think there will be multiple clinical infertility solutions involving gene therapy soon."

How soon?

"The next couple years," he says. "It would be very hard to resist."

In his talk at the National Academies meeting, Orwig flashed a slide that said: "Germ-Line Gene Therapy Is Technically Feasible Today." Afterward, according to Orwig, a member of the planning committee sidled up to him backstage and said, "Germ-line gene therapy is going to happen, I guarantee it." That sentiment never made it into the final meeting communiqué. But it galvanized Orwig.

"Whereas it might have been something I wanted to do hiding under a rock, it's now like, 'You go, man!' Let me get to work and prove to you that I can do it."

In animals, of course.

A Gentle Nudge Down the Slope

A few steps down the hallway from Orwig's office is a complex of rooms that houses hundreds of mice. You have to don a gown and booties and a mask to enter–not because you might catch something from the mice but because they might catch something from you. Many of the cages contain what are called nude mice–pink, wrinkled little rodents that resemble scrotums with eyes and feet. They are nude in the sense that they were bred to have compromised immune systems that accommodate cells transplanted from other species–for instance, human spermatogonial stem cells with mutations–to allow researchers to better understand the biology of male infertility.

If, as Church says, "everything is going to be done in animals first," the road to human germ-line modification runs through rooms like these. CRISPR makes the task more efficient ("It is so freaking easy!" Orwig says), but scientists have been able to alter the genes of sperm-making cells for more than two decades, beginning in 1994, when University of Pennsylvania biologist Ralph Brinster (Orwig's mentor) did the pioneering experiments in mice.

Male infertility has many causes, including obstructive "plumbing" issues, glitches in the incredibly complex process of sperm creation, and underachieving sperm. But in many cases, males simply can't make sperm at all; the condition, known as nonobstructive azoospermia, affects roughly 350,000 men in the U.S., according to Orwig. Several genes have been associated with the failure to produce sperm, including *tex11* and *sohlh1*, and those cases form the backdrop of the experiment Orwig is eager to do.

What Orwig wants to do is take infertile mice, which have a dysfunctional version of one of these genes, remove the sperm-forming stem cells from their testes and correct the defect in those cells by using the new gene-editing techniques. Once the altered stem cells are grown to sufficient numbers in the test tube and screened for precisely the correct alteration, they can be transplanted back into the testes of the animals. And at least in animal experiments of this kind, there is no need for any fancy molecular tests–if the gene editing is successful, Orwig will know within a couple of months because the infertile males will unambiguously demonstrate their ability to become fathers.

"We've been transplanting stem cells for 25 years in almost every species–mice, rats, hamsters, sheep, goats, pigs, dogs and monkeys," Orwig says. "That's a pretty broad swath of evolution, and in all this time, in all these animals, as far as we know, nothing bad has happened." That is why Orwig is optimistic he can demonstrate that editing the genes of stem cells in mice can reverse infertility.

This may seem like an innocuous animal experiment, but to edit a sperm-forming stem cell is to permanently modify the germ line because the resulting sperm cells pass the correction along to the next generation. A potential treatment for male infertility would cross the red line. And although Orwig has no plans to do the obvious human follow-up in his Pittsburgh lab, a successful preclinical demonstration in mice and primates would provide the impetus for an attempt in the private sector–which is where Church believes the final steps will unfold. "Sperm-editing efforts will be privately funded," he says, "just like other therapies."

Developing such a clinical treatment would face technical hurdles, of course. For one thing, scientists would have to find a way to maintain human spermatogonial stem cells long enough to select the right ones for transplantation–still not a trivial task. But these male stem cells offer much less of a moving target than embryos, which are dynamic and change rapidly. The Chinese researchers who have attempted gene editing in embryos with CRISPR, for example, have reported both "untoward mutations" and "mosaicism," meaning that some cells in the embryos show successful editing, whereas others do not. Moreover, the DNA of gene-edited stem cells can be screened before an embryo is produced.

That is what makes Orwig's potential mouse experiment so politically inconvenient. Because of prohibitions enacted by Congress in the 1990s, the National Institutes of Health cannot fund any research that involves the destruction of human embryos. A human version of Orwig's proposed mouse experiment might sidestep that prohibition, but it would probably fall under a new obstacle that the House of Representatives introduced two weeks after the December summit on gene editing. In a two-sentence passage buried in the 2,009-page omnibus spending bill of 2015, Congress inserted language forbidding the FDA to consider any medical intervention relying on the use of gene-edited embryos; the wording does not explicitly prohibit editing germ cells, but Stanford University law professor Henry Greely believes "the FDA would take the position that those sperm were more than minimally manipulated human cells that would require FDA approval as a drug or biological product." The regulatory piece, he thinks, could add a decade or two to Church's timeline.

That does not mean Orwig's mouse experiment would be against the law—just a gentle nudge down the slippery slope toward germ-line modification. The step across the red line could happen in private IVF clinics, which have a long (and blemished) history of pushing the envelope on new reproductive technologies. "It's such an easy technology to apply that it would only take somebody with a little chutzpah to get together with someone in an IVF clinic and, you know, take a shot at it," says George Daley, a stem cell biologist at Boston Children's Hospital. "This is coming down the pike, and people need to start thinking about it," he notes. "This is a potentially disruptive reproductive technology."

It probably will not happen in the U.S. unless the public—and political—perception of germ-line modification becomes more accommodating, but Orwig is quietly preparing for that day. "We're going to work real hard behind the scenes," he says, "until the worldview changes."

Crossing Borders

The "worldview" on germ-line editing is complicated and contradictory. A majority of Americans do not like the idea of editing genes in either embryos or germ cells, according to a recent analysis of 17 public opinion polls published in the *New England Journal of Medicine*. Yet paradoxically, most people support gene editing in adults "aimed at preventing one's children from inheriting certain diseases." (Robert J. Blendon, lead author of the study, says any intervention on the adult side that is positive for the next generation, including germ cells, would have "considerable public support.") Moreover, the *NEJM* study also pointed out that many of these public opinion polls pose their questions using language that "might not be scientifically precise." In other words, although the National Academies meeting adjourned last December with a pledge to continue the public conversation about germ-line editing, it is not clear that the public even understands the terms of that conversation. And while public forums struggle to find an effective vocabulary, the science races ahead.

As we spoke in his office last spring, Orwig nodded at a scientific reprint sitting on his desk. "I really, really love this paper," he said. He was referring to research published this past February in the journal *Cell Stem Cell* by a group headed by Qi Zhou of the Chinese Academy of Sciences. The experiment basically provided a recipe for the in vitro creation of germ cells.

The researchers showed that they could create sperm-forming stem cells in a dish; with a technique currently used in IVF clinics, these cells could be injected into egg cells to create fertile male mice. Harvard's Daley says of this advance: "With the addition of CRISPR, you've got the brave new world."

When Aldous Huxley imagined his brave new world in 1932, the story unfolded under one totalitarian regime, with neither national boundaries nor local regulations. In today's world, germ-line editing in any one place means the germ line is edited everywhere. "Regulation is country-specific, but science crosses borders," Harvard Law's Cohen says. Even if there were laws against germ-line modification in the U.S., you would have to build a wall much higher than the one proposed by Donald Trump to keep American germ lines insulated from an eventual influx of modified DNA.

"If you play out the world to 100 years from now, if anybody does this anywhere, that's the end game," Cohen says. "Over time those people will mate and create offspring and will cross borders and enter our shores. And if the safety and efficacy are worked out, it's inevitable that you're going to have people walking around in the world, and they're going to be reproducing, and they will end up in this country, and those changes will enter the U.S. gene pool."

As I am concluding my visit with Orwig, he glances at the computer on his desk. A reporter has sent an e-mail seeking comment on yet another experiment cozying up to the red line: a group in China has just reported its attempt to edit human embryos (nonviable) to be resistant to HIV infection. "Eventually we'll learn a vocabulary that acknowledges that we're there," Orwig says. "But I feel we're already there.

Referenced

The Pandora's Box Congress. Michael Rogers in *Rolling Stone*; June 19, 1975.

CRISPR Germline Engineering–The Community Speaks. Katrine S. Bosley et al. in *Nature Biotechnology*, Vol. 33, pages 478–486; May 2015.

Experimental Methods to Preserve Male Fertility and Treat Male Factor Infertility. Kathrin Gassei and Kyle E. Orwig in *Fertility and Sterility*, Vol. 105, No. 2, pages 256–266; February 2016.

Complete Meiosis from Embryonic Stem Cell-Derived Germ Cells in Vitro. Quan Zhou et al. in *Cell Stem Cell*, Vol. 18, No. 3, pages 330–340; March 3, 2016.

About the Author

Stephen S. Hall is an award-winning science writer and regular contributor. He is author, most recently, of Wisdom: From Philosophy to Neuroscience *(Knopf, 2010).*

A New Era of Designer Babies May Be Based on Overhyped Science

By Laura Hercher

For better or worse, genetic testing of embryos offers a potential gateway into a new era of human control over reproduction. Couples at risk of having a child with a severe or life-limiting disease such as cystic fibrosis or Duchenne muscular dystrophy have used preimplantation genetic testing (PGT) for decades to select among embryos created through in vitro fertilization (IVF) for those that do not carry the disease-causing gene. But what new iteration of genetic testing could tempt healthy, fertile couples to reject our traditional time-tested and wildly popular process of baby making in favor of hormone shots, egg extractions and DNA analysis?

A California-based start-up called Orchid Biosciences claims it has an answer to that question. The company offers prospective parents genetic testing prior to conception to calculate risk scores estimating their own likelihood of confronting common illnesses such as heart disease, diabetes, and schizophrenia and the likelihood that they will pass such risks along to a future child. Parents-to-be can then use IVF, along with Orchid's upcoming embryo screening package, to identify the healthiest of their embryos for a pregnancy.

Orchid aims to use PGT and IVF to expand what is already a thriving marketplace in screening tests for prospective parents. Initially, the only people offered tests to prevent genetic disease in the next generation were those whose ancestry put them at higher risk for a specific condition, such as Tay-Sachs disease in the Ashkenazi Jewish population. The first genetic screen intended for universal use and covering a wide range of diseases was introduced by Counsyl (now part of Myriad Genetics) in 2010. Today carrier screening is a $1.7-billion industry. These tests search for genetic problems that otherwise come to light only after the birth of an affected child. But diseases caused by a single gene are rare. Most children are born healthy, and most couples who do carrier screening come away reassured.

By contrast, Orchid's risk assessment includes common diseases, ensuring that a high percentage of prospective parents who do this version of preconception testing will find something to worry about. Those who choose to act on their concerns will soon have the option of paying for IVF plus Orchid's embryo-testing package. According to its promotional materials, the company will provide a scorecard intended to identify, among various embryos, the future children least likely to develop heart disease, breast cancer, prostate cancer, type 1 or 2 diabetes, and five other conditions that make up Orchid's current common disease risk portfolio.

With a marketing strategy that encourages routine use of IVF for those who can afford it, Orchid breaks new ground in introducing the first—but likely not the last—consumer-driven model of human reproduction. The ambitions of this new Silicon Valley venture into health care are backed by the imprimatur of health-tech luminaries, including 23andMe co-founder and Orchid investor Anne Wojcicki. Orchid's first product on the market is its "Couple Report," at a cost of $1,100. Phase two, scheduled for launch later this year, examines embryos conceived by IVF, allowing the couple to pick and choose among potential children in a process that Orchid CEO Noor Siddiqui, speaking in an interview on the podcast Mendelspod in April, referred to as "embryo prioritization." Siddiqui is a former Thiel Foundation Fellow whose interests lie in the use of technology in medicine. She did not respond to repeated requests for an interview from *Scientific American*.

Geneticists have greeted Orchid's launch with skepticism, in large part because of objections to the company's use of a technique called polygenic risk scores to assess an embryo's lifetime risk of common diseases. Heart disease runs in families just like musical ability or height, but only in exceptional cases can the inherited risk be traced to a single gene. Hundreds or even thousands of genes each contribute in a small way. Polygenic risk scores attempt to sum up the overall likelihood of a particular outcome—such as getting a disease–by simply observing which patterns of variation in a genome are associated with a higher or lower probability of having the condition. In other words, this method gives us information about who might be more or less likely to get sick without explaining why. The statistical association is real but hardly definitive, and it tracks population-level trends that may not be relevant for the individual in question.

Researchers who work with polygenic risk scores are concerned about their use in this context. "We don't know what these variants are doing biologically," says Peter Kraft, a professor of epidemiology and biostatistics at the Harvard T. H. Chan School of Public Health. "Something that's associated with a decreased risk of breast cancer could be associated with all other kinds of things, some of which might actually increase your risk of something else. We just don't know enough yet."

Some version of prenatal planning as envisioned by Orchid may be possible eventually, but few experts seem to share their optimism that today is that day. A July 1 special report in the *New England Journal of Medicine* pointed out the inherent weakness of using polygenic risk scores to distinguish among sibling embryos–which, unlike random individuals in a population, will be identical in 50 percent of the genetic variation that is examined to generate a score. The report concluded with recommendations on how to convey any purported benefits from polygenic scores in embryo selection responsibly–and the need to emphasize the underlying uncertainties in the data. "Any one of the issues discussed in this article would be difficult to communicate accurately–even to other scientists and clinicians," the authors noted. "Collectively, these issues constitute a formidable challenge for [companies selling these services], which must ensure that their customers understand what they are doing." The report also called for the Federal Trade Commission to look carefully at claims made by any company using polygenic scoring to pick embryos.

Current polygenic risk scores have limited predictive strength and reflect the shortcomings of genetic databases, which are overwhelmingly Eurocentric. Alicia Martin, an instructor at Massachusetts General Hospital and the Broad Institute of the Massachusetts Institute of Technology and Harvard University, says her research examining polygenic risk scores suggests "they don't transfer well to other populations that have been understudied." In fact, the National Institutes of Health announced in mid-June that it will be giving out $38 million in grants over five years to find ways to enhance disease prediction in diverse populations using polygenic risk scores. Speaking of Orchid, Martin says, "I think it is premature to try to roll this out."

In an interview about embryo screening and ethics featured on the company's Web site, Jonathan Anomaly, a University of Pennsylvania bioethicist, suggested the current biases are a problem to be solved by getting customers and doing the testing. "As I understand it," he said, "Orchid is actively building statistical models to improve ancestry adaptation and adjustments for genetic risk scores, which will increase accessibility of the product to all individuals."

Still, better data sets will not allay all concerns about embryo selection. The combined expense of testing and IVF means that unequal access to these technologies will continue to be an issue. In her Mendelspod interview, Siddiqui insisted, "We think that everyone who wants to have a baby should be able to, and we want our technology to be as accessible to everyone who wants it," adding that the lack of insurance coverage for IVF is a major problem that needs to be addressed in the U.S.

But should insurance companies pay for fertile couples to embryo-shop? This issue is complicated, especially in light of the fact that polygenic risk scores can generate predictions for more than just heart disease and cancer. They can be devised for any trait with a heritable component, and existing models offer predictions for educational attainment, neuroticism and same-sex sexual behavior, all with the same caveats and limitations as Orchid's current tests for major diseases. To be clear, tests for these behavioral traits are not part of Orchid's current genetic panel. But when talking about tests the company does offer, Siddiqui suggested that the ultimate decision makers should be the parents-to-be. "I think at the end of the day, you have to respect patient autonomy," she said.

Despite Orchid's hard lean into parental free choice, bioethicists such as Gabriel Lázaro-Muñoz of the Center for Medical Ethics and Health Policy at Baylor College of Medicine worry that Orchid's system of ranking embryos may unduly influence prospective parents and replace a very necessary broader societal debate on what qualifies as a good life. It is problematic for that reason, according to Lázaro-Muñoz, to have these companies "bias the conversation."

Lurking in the background of every discussion on embryo selection and ethics is the specter of eugenics. "I think we have to be very aware of our history," Lázaro-Muñoz says, "in terms of sterilization and state-mandated programs in the past that were aimed at ... exterminating individuals with some of these conditions."

Clearly Orchid anticipates pushback. The company's promotional materials include guides to fertility planning and the genetics of irritable bowel disease but also a set of talking points for concerned relatives described as "How to respond to your family skeptics–playing God, designer babies, and genetic enhancement."

"Yes, we're going there," the guide says. Ethics? Bring it on. This is not a company in a defensive crouch. The "Our Principles" section of its Web site positions genetic testing as a human right. "From a reproductive freedom perspective, we stand for a couple's right to have access to information that enables them to mitigate disease risk for their future child," it says. Like other Silicon Valley health care technology pioneers, Orchid presents itself less as a product than as a social justice movement with a little commercial venture on the side, like a gift shop.

Orchid dismisses suggestions from detractors that its marketing oversells what polygenic risk scores have to offer in the context of screening embryos. "Parents are asking for this information and deserve to know it," Siddiqui told Mendelspod, warning that those who stand in the way are "frankly being a little bit paternalistic." And if prospective parents are not asking, Siddiqui suggested, perhaps they should be. When it comes to the next generation, "we've been sort of just rolling the dice," Siddiqui said, while "the ability to actually stack the odds against disease is ... sort of a new capability that humanity has just gotten online."

The suggestion that embryo selection is not only something people can do but something they *should* do raises perhaps one of the thorniest ethical issues of all. In the Mendelspod interview, Siddiqui drew a contrast between "earned" and "unearned" bad luck. "You can get hit by a car, right? That's totally out of your control. But what is earned bad luck?" she asked before answering her own question. "I mean, that's the idea of ... you're going base jumping constantly, and then you break your leg.... You kind of exposed yourself to higher risk there."

Ultimately, if technology allows Orchid to offer a product that meaningfully reduces the risk of disease susceptibility in the next generation, does that mean that anyone who can't or won't use it deserves their bad luck? If the basic, no-frills version of human reproduction comes to be seen as a form of careless parenting, it invites a callous parsing of who does and does not deserve their fate—and, by extension, who does and does not deserve resources and support.

About the Author

Laura Hercher is a genetic counselor and director of student research at the Joan H. Marks Graduate Program in Human Genetics at Sarah Lawrence College. She has written broadly on ethical, legal and social issues related to genetic medicine. Hercher is the host of The Beagle Has Landed, *a podcast for the clinical genetics community and other sci-curious individuals.*

Fighting Unfairness in Genetic Medicine

By Stephanie Devaney

When the race to sequence the first human genome was rushing toward the finish line about 20 years ago, I remember feeling mesmerized by what was about to happen. It was the dawn of a new century, and it seemed we were on the cusp of unlocking the meaning behind the blueprint of life, DNA. Once we could line up all 3.1 billion base pairs of the molecule in our genome, I thought—I was an undergraduate student at the time, dazzled by science—we would understand everything there is to know about human health and disease.

What I didn't know was that those first decades of genetic medicine would leave a lot of people behind. So I was taken aback several years later, in 2009, just after I got my doctorate in molecular genetics, when researchers at Duke University reported that 96 percent of the genomic data we had gathered came from people of European ancestry. This was not the result of small numbers: they calculated the percentage using the more than 1.7 million individual genome samples analyzed at the time, but the samples were lacking diversity. Over the next few years things did not get much better, and as recently as four years ago genomic databases were still way out of balance, with more representation of Europeans and less of everyone else.

This inequity, if it is not fixed, will turn into tremendous health inequality. Today more and more people are getting answers about the underlying causes of their diseases because of medicine's ability to mine their genomes. There are hundreds of drugs that contain genetic information in their labeling because gene variants affect how bodies process these drugs, and knowing the variants that patients have helps doctors set the most beneficial dose for their patients. Moreover, today improved knowledge about the genomic drivers of different cancers has paid dividends in how physicians diagnose and treat many tumors. Yet people who are not white and not male have different sets of genes that do not always fit into these treatment regimens.

For example, African-Americans and Latinos have the highest rate of asthma in the U.S., but studies show that common drugs used in inhalers do not help them as well as they help whites. Asians who take the antiseizure drug carbamazepine have a higher risk of a severe, sometimes fatal, reaction. Nobody developing these drugs, or prescribing them when they first came into use, anticipated these problems. If DNA is one important factor in our quest for more effective medical treatment, we need to address the lack of diversity in genetic data.

That is where the *All of Us* Research Program, where I work, hopes to help. Set up by the National Institutes of Health and launched in 2018, we are asking a million or more people from all backgrounds to join us as partners in research, not as human subjects, and share all kinds of health information over the course of their lives. Already we have more than 250,000 participants. More than 51 percent belong to racial and ethnic minorities, more than 10 percent are sexual and gender minorities, and overall more than 80 percent represent a group that has been historically underrepresented in research data sets.

People can join All of Us by going to our program Web site (www.joinallofus.org) and clicking "Join Now." After agreeing to participate, respondents can offer us their medical records, answer a variety of surveys about their health and lifestyle, and participate in other activities such as syncing their fitness tracker data to our program. We also have hundreds of enrollment sites at local hospitals and health centers across the country where participants can provide samples of blood and urine to help researchers study their DNA. Our hope is for people to stick with us for 10 years or more because, as the program grows, we will regularly add new ways for them to learn about themselves and contribute to research.

The Moment is Right

A lot of this participant-researcher collaboration is linked to advances in technology. Sequencing that first human genome had a $1-billion price tag. Today such a sequence costs less than $1,000 and can take less than 24 hours to complete. It is also easier to integrate this information with other crucial medical data. Health care organizations have been turning their patients' paper-based medical records into electronic versions. As of 2017, 96 percent of all U.S. hospitals and 80 percent of all office-based doctors are using a certified electronic health record system. New apps on smartphones and other digital health technologies such as smart watches collect data from nearly anywhere and directly from a person. These trends all make it easier to store, share and mine large data sets for answers to questions about disease causes and effects. Such trends also raise big and disturbing issues about privacy, making it important for projects such as ours to have both strong security and full transparency to all our participants.

And it is crucial to treat these people as partners. The actions of past medical researchers have earned much distrust in minority communities, after causing harm in the Tuskegee Syphilis Study, where researchers misled African-American men with syphilis and never gave them adequate treatment, and with the widespread use of HeLa cells, which were taken from a patient named Henrietta Lacks without her knowledge or permission. People wanted to see research go forward but *with* them rather than about them. To overcome this kind of distrust, All of Us is using a new model for research, one that invites input from participants as well as researchers with science degrees. Participants serve on the program's advisory and governing bodies, working groups, and task forces. We have also partnered with local health care organizations, hospitals, and community groups to advise us and help find people to participate. Community engagement is not familiar ground for large medical research projects, and we are still learning the best ways to do it.

Some studies have provided us with blueprints for developing long-term relationships like the ones we hope to have, studies that have changed medicine for the better. The Framingham Heart Study, for example, started in 1948 with 5,209 men and women, largely white, from one town in Massachusetts. With a 99 percent retention rate, the study continues to this day. As participants share data year after year, researchers can see how their heart health changes over time. The risk factors for heart disease identified by the Framingham study–such as high blood pressure, high cholesterol, smoking and obesity–are so ingrained in our collective consciousness and our approach to health care that they feel like common sense.

Going Further

This kind of medical discovery is what we envision for All of Us, but we want to take it further, with participants who are not all white and who represent diversity in many dimensions, not just traditional race labels that, in reality, encompass a lot of different backgrounds. If we're going to get at the root causes of health and disease, this means understanding the differences and similarities among us all. For example, sickle cell disease occurs when someone inherits two mutated genes for the oxygen-carrying protein hemoglobin. It affects 100,000 African-Americans and more than 20 million people around the world. In contrast, sickle cell trait–meaning just one of these genes is mutated–actually gives people an advantage in surviving malaria, which makes evolutionary sense if your ancestors came from areas such as Africa where malaria is prevalent. New studies, however, have found that sickle cell trait might not be as benign as doctors used to believe, because it may increase the risk for kidney disease. Some African-Americans are more susceptible to this risk and some less. There's clearly more to learn about why this might be the case and about how different DNA variants might interact to affect the health of people with sickle cell trait. The DNA information from more than a million All of Us participants could help researchers learn much more about complex traits like this.

We do have to start with some of the broad-brush categories to recruit enough people to start recognizing the more fine-grained groups among them. Currently we are exceeding our goal of overrepresenting groups that have been historically underrepresented in research. For instance, African-Americans make up about 13 percent of the U.S. population but just 3 percent of the samples previously used in genome studies. In All of Us, 21.5 percent of participants so far are African-American. Similarly, Hispanics constitute about 18 percent of the U.S. population but in 2016 made up less than 1 percent of the data in our genomic databases. Today 17.6 percent of All of Us participants are Hispanic.

That diversity will help us discover more about how DNA affects health across different communities, but the molecule will not be our sole focus. Many factors beyond our genes are at play when it comes to disease. We know that where you were born, what you eat, the stress you feel, and other clinical and biological factors affect health, but we still don't understand by how much. For example, when we think about some of the most common chronic diseases that afflict our population–high blood pressure is one example–many of them disproportionately affect the most socially and economically disadvantaged people in our country. And from what we can tell at the moment, the determinants are not simply their race or ethnicity. Risks also include family structure, socioeconomic status, stressors such as trauma, sex and gender inequality, availability of nutrient-rich foods, access to health care, and many other factors that we can capture in the All of Us data set.

Within the next several years, we should be able to compare this rich set of information with participants' DNA. When we do so, scientists such as myself, the All of Us participants and all of you will start to get a clearer picture of the roles that biology and environment play in disease development, and–most important of all–what we can do about it.

About the Author

Molecular geneticist Stephanie Devaney is deputy director of the All of Us research program at the National Institutes of Health. She was the staff lead for the White House Precision Medicine initiative.

A Vision for the Next Decade of Human Genomics Research

By Eric D. Green

Since the launch of the Human Genome Project 30 years ago, genomics has become deeply woven into research, medicine and, increasingly, daily life. In roughly 10,000 days, genomics has blossomed from a boutique discipline associated with a moonshot project into a scientific area now vital for areas as diverse as microbiology and oncology.

Numerous genomic applications—noninvasive prenatal genetic testing, DNA-based forensics, direct-to-consumer ancestry sleuthing, genetic disease diagnostics and SARS-CoV-2 virus surveillance, among others—are common practice.

What's next for genomics, especially in its applications to human health and disease? For the past several years, the broader genomics community has worked with the U.S. National Human Genome Research Institute to formulate a vision—a "2020 vision"—of what lies at the forefront of human genomics for the next decade.

The vision represents the strong consensus that we heard about the key guiding principles and values that now undergird human genomics, such as those related to diversity, equity, social justice, data sharing and team science. The broader biomedical enterprise increasingly depends on a strong foundation for genomics (consisting of infrastructure, resources and dynamic areas of technology development), and so the vision articulates the need to sustain and improve this foundation.

The new vision identifies barriers that impede progress in genomics—with a call to action for high-risk pursuits to break down such barriers. And, because we are scientists, there are also ambitious and compelling research projects in human genomics that serve to stretch our imaginations and pique our curiosity.

The vision for the next phase of human genomics is bolder than ever, and continues to embrace many of the original promises of the Human Genome Project. It does so even as some of the previous predictions about genomic advances are still to be fully realized.

Critics may say such genomics boldness is disingenuous and breeds overpromises, but at every encounter during our recent strategic planning process, we found the optimism and exuberance of our colleagues to be both inspiring and metaphorically intoxicating, pushing us to be even bolder and more willing to take risks in describing where genomics might lead us. That is why the vision's capstone is a set of 10 bold predictions for human genomics by 2030.

Bold Predictions For Human Genomics by 2030

1. Generating and analyzing a complete human genome sequence will be routine for any research laboratory, becoming as straightforward as carrying out a DNA purification.
2. The biological function(s) of every human gene will be known; for non-coding elements in the human genome, such knowledge will be the rule rather than the exception.
3. The general features of the epigenetic landscape and transcriptional output will be routinely incorporated into predictive models of the effect of genotype on phenotype.
4. Research in human genomics will have moved beyond population descriptors based on historic social constructs such as race.
5. Studies that involve analyses of genome sequences and associated phenotypic information for millions of human participants will be regularly featured at school science fairs.

6. The regular use of genomic information will have transitioned from boutique to mainstream in all clinical settings, making genomic testing as routine as complete blood counts.
7. The clinical relevance of all encountered genomic variants will be readily predictable, rendering the diagnostic designation 'variant of uncertain significance (VUS)' obsolete.
8. An individual's complete genome sequence along with informative annotations will, if desired, be securely and readily accessible on their smartphone.
9. Individuals from ancestrally diverse backgrounds will benefit equitably from advances in human genomics.
10. Breakthrough discoveries will lead to curative therapies involving genomic modifications for dozens of genetic diseases.

From Green, E.D., et. al., "Strategic vision for improving human health at the Forefront of Genomics," Nature 2020.

Innovation is inspired by reaching for seemingly unreachable goals. When early naysayers predicted that the Human Genome Project would fail, we ignored the pessimism and generated the first human genome sequence in less time and for less money than originally planned. Had we not been overly ambitious for the past 17 years since the end of the Human Genome Project, we would not have reduced the cost of DNA sequencing by more than a millionfold, made genome sequencing routine for the diagnosis of rare genetic diseases, or, most recently, generated a truly complete, end-to-end sequence of a human chromosome. As a physician-scientist, I was drawn to genomics because of its potential for changing medical care, never dreaming that genomic medicine would begin to be implemented in my lifetime (and I'm only 60!).

During the Human Genome Project, we mapped and sequenced the human genome to the best of our ability with the then-available methods, which shockingly started with handwriting DNA sequences onto notebook pages and then faxing those pages to collaborators. Little by little, we found that our step-by-step evolution of approaches was in fact the revolution that we needed for success.

Similar patience and tenacity will be needed in striving for the vision's bold predictions. Even if two or three of them are realized, we will have accomplished the incredible.

Among the boldest of the 10 predictions are those that rely not only on the tools of genomics, but also on advances in society—a reminder that the science can only take us so far. To fulfill the full potential of genomics, researchers need to push the boundaries of science while every one of us—scientists and nonscientists alike—need to continually examine what we know to be true and know to be possible. Each of us must raise awareness of health inequities and develop the empathetic will to address them, caring for others' genomes as much as we do our own.

About the Author

Eric D. Green is director of the National Human Genome Research Institute (NHGRI) at the National Institutes of Health, a position he has held since 2009. NHGRI is the largest organization in the world solely dedicated to genomics research.

GLOSSARY

antigen A harmful substance that causes a body to produce antibodies.

biotechnology The use of living cells to make useful products.

clone A plant or animal that is grown from one cell of its parent and that has exactly the same genes as its parent.

conservation The protection of animals, plants, and natural resources.

enzyme A chemical substance in animals and plants that helps to cause natural processes.

gene A part of a cell that controls or influences the appearance, growth, etc., of a living thing.

heterogeneous Made up of parts that are different.

hybridization The process of breeding two plants or animals of different species.

immunodeficiency The inability to produce antibodies.

immunotherapy Treatment or prevention of a disease by stimulating, enhancing, or suppressing the immune system.

immutable Unable to be changed.

infertility Not able to reproduce.

invasive Tending to spread.

metabolic Relating to the processes by which substances are handled by a living body.

molt To lose a covering of hair, skin, or feathers and have it replaced by new growth.

mutation A change in the genes of a plant or animal that causes physical changes that are not normal.

pathogen Something that causes disease.

RNA A part of a cell that controls or influences the appearance, growth, etc., of a living thing.

socioeconomic Related to or involving a combination of social and economic factors.

stressors Things that make you worried or anxious.

FURTHER INFORMATION

"Busting the Myths on New GMOs," *Slow Food*, November 16, 2022, https://www.slowfood.com/busting-the-myths-on-new-gmos/.

"WGS More Effective than Targeted Gene Sequencing For Diagnosing Genetic Diseases in Newborns," *Genetic Engineering & Biotechnology News*, July 11, 2023, https://www.genengnews.com/topics/translational-medicine/wgs-more-effective-than-targeted-gene-sequencing-for-diagnosing-genetic-diseases-in-newborns/.

Cobb, Matthew. "Gene Drives Could Fight Malaria and Other Global Killers but Might Have Unintended Consequences," *Scientific American*, January 13, 2023, https://www.scientificamerican.com/article/gene-drives-could-fight-malaria-and-other-global-killers-but-might-have-unintended-consequences/.

Mullin, Emily. "A New Cloned Horse Offers Hope for Endangered Species," *Wired*, May 4, 2023, https://www.wired.com/story/cloning-endangered-species-przewalskis-horse/.

Murugesu, Jason Arunn. "People with total colour blindness able to see red after gene therapy," *NewScientist*, July 10, 2023, https://www.newscientist.com/article/2381757-people-with-total-colour-blindness-able-to-see-red-after-gene-therapy/.

Rossi, Melissa. "As climate change threatens European agriculture, debate over GMO crops is reignited," *Yahoo! News*, February 21, 2023, https://news.yahoo.com/as-climate-change-threatens-european-agriculture-debate-over-gmo-crops-is-reignited-195549518.html.

Spahich, Niki. "The Future of Gene Therapy for a Rare Pediatric Autoimmune Disease," *TheScientist*, July 10, 2023, https://www.the-scientist.com/news/the-future-of-gene-therapy-for-a-rare-pediatric-autoimmune-disease-71214.

Wickelgren, Ingrid. "Bacterial 'Nanosyringe' Could Deliver Gene Therapy to Human Cells," *Scientific American*, March 29, 2023, https://www.scientificamerican.com/article/bacterial-nanosyringe-could-deliver-gene-therapy-to-human-cells/.

CITATIONS

1.1 Gene Therapy's Second Act by Ricki Lewis (March 1, 2014); 1.2 How RNA Discoveries Are Radically Changing Gene Therapy and Other Medical Treatments by Christine Gorman, Dina Fine Maron (March 18, 2014); 1.3 What Gene Therapy Needs Now: A Good Off Switch by Jim Kozubek (January 1, 2016); 1.4 Mysterious 'Retron' DNA Helps Scientists Edit Human Genes by Saugat Bolakhe on May 1, 2022; 1.5 Could the Next Big Information Technology Be ... DNA? by James E. Dahlman (June 1, 2019); 2.1 CRISPR Gene Editing May Help Scale Up Coronavirus Testing by Jim Daley (April 23, 2020); 2.2 How Designer DNA Is Changing Medicine by Carolyn Barber (July 17, 2021); 2.3 The Definition of Gene Therapy Has Changed by Esther Landhuis (November 1, 2021); 2.4 Four Success Stories in Gene Therapy by Jim Daley (November 1, 2021); 3.1 New Gene-Editing Techniques Could Transform Food Crops--or Die on the Vine by Stephen S. Hall (March 1, 2016); 3.2 Tweaking Vegetables' Genes Could Make Them Tastier–And You'll Get to Try Them Soon by Meghan Bartels (May 10, 2023); 3.3 Why People Oppose GMOs Even Though Science Says They Are Safe by Stefaan Blancke (August 18, 2015); 3.4 Stop Arguing over GMO Crops by Sarah Garland (December 14, 2020); 4.1 The American Chestnut's Genetic Rebirth by William Powell (March 1, 2014); 4.2 Could Genetic Engineering Save the Galapagos? by Stephen S. Hall (November 1, 2017); 4.3 De-extinction Company Aims to Resurrect the Tasmanian Tiger by Kate Evans (August 16, 2022); 4.4 To Restore Biodiversity, Embrace Biotech's 'Intended Consequences' by Paul Robbins, Ryan Phelan on June 15, 2021; 5.1 The First Tinkering with Human Heredity May Happen in the Infertility Clinic by Stephen S. Hall (September 1, 2016); 5.2 A New Era of Designer Babies May Be Based on Overhyped Science by Laura Hercher (July 12, 2021); 5.3 Fighting Unfairness in Genetic Medicine by Stephanie Devaney (January 1, 2020); 5.4 A Vision for the Next Decade of Human Genomics Research by Eric D. Green (October 28, 2020

Each author biography was accurate at the time the article was originally published.

INDEX

D

E

F

G